Oxford Resources for Cambridge

Core

AF669841

Cambridge IGCSE®
Complete
Mathematics

Sixth Edition

Teacher Handbook

Ian Bettison
Mathew Taylor

OXFORD
UNIVERSITY PRESS

OXFORD
UNIVERSITY PRESS

Great Clarendon Street, Oxford, OX2 6DP, United Kingdom

Oxford University Press is a department of the University of Oxford. It furthers the University's objective of excellence in research, scholarship, and education by publishing worldwide. Oxford is a registered trade mark of Oxford University Press in the UK and in certain other countries.

© Oxford University Press 2024

The moral rights of the authors have been asserted

First published in 2024

All rights reserved. No part of this publication may be reproduced, stored in a retrieval system, or transmitted, used for text and data mining, or used for training artificial intelligence, in any form or by any means, without the prior permission in writing of Oxford University Press, or as expressly permitted by law, by licence or under terms agreed with the appropriate reprographics rights organization. Enquiries concerning reproduction outside the scope of the above should be sent to the Rights Department, Oxford University Press, at the address above.

You must not circulate this work in any other form and you must impose this same condition on any acquirer

British Library Cataloguing in Publication Data

Data available

9781382042512

10 9 8 7 6 5 4 3 2 1

Paper used in the production of this book is a natural, recyclable product made from wood grown in sustainable forests.

The manufacturing process conforms to the environmental regulations of the country of origin.

Printed in Great Britain by Ashford Colour Press Ltd

Acknowledgements

The publisher would like to thank the following for permissions to use copyright material:

Artwork by Aptara Inc., Thompson Digital, and Oxford University Press.

Cover: Dimitri Otis / Photodisc / Getty Images.

Although we have made every effort to trace and contact all copyright holders before publication this has not been possible in all cases. If notified, the publisher will rectify any errors or omissions at the earliest opportunity.

FSC www.fsc.org MIX Paper | Supporting responsible forestry FSC® C011748

Contents

Introduction

Welcome to OUP's *Cambridge IGCSE® Complete Mathematics Core: Teacher Handbook Sixth Edition.* This comprehensive guide is designed to support you as educators in delivering an enriching and effective learning experience for students pursuing the Cambridge IGCSE® in Mathematics.

As educators, you play a pivotal role in shaping the mathematical understanding and skills of your students. This handbook has been designed to provide you with practical activities and support resources that align with the *Cambridge IGCSE® Complete Mathematics Core: Student Book Sixth Edition.* Whether you are a seasoned mathematics teacher or new to the Cambridge curriculum, this handbook is a valuable companion on your teaching journey.

Content overview

Objectives for each chapter are listed and the relevant chapter(s) of both the Student Book and this handbook are mapped to each objective. This will enable you to quickly find the relevant teaching material, whatever programme of study you choose to adopt for delivering the material to your students.

Chapter contents

Each chapter of this handbook is presented in the same format and with the same set of useful resources.

- **Chapter introduction** The introduction provides important information on the purpose of the chapter, a summary of the contents and prerequisite knowledge.
- **Investigations, activities and puzzles** These activities are designed to supplement the teaching of the syllabus content and provide students with opportunities for enrichment. Each activity has teaching notes to help you plan and deliver the activity as well as suggestions for extending the activities further.
- **Glossary** It is important that students understand the key terminology used in mathematics. These are listed, along with comprehensive definitions, in the glossary for each chapter.
- **Extension worksheet** More advanced students will make quicker progress through the contents of each chapter and we have provided an extension worksheet to enable these students to be stretched beyond the limits of the syllabus, while continuing to do relevant mathematics.

Revision checklists

Each of the chapters in the book covers many aspects of the Cambridge IGCSE® syllabus. The revision checklists provide students with a summary list, in student-friendly language, to help them revise for the end-of-course examinations.

Each checklist can be copied and issued to the students, either at the end of the course or at an appropriate point during the course.

Students tick to indicate that they have revised the topics and whether they feel secure with their knowledge. This gives them responsibility for their learning and a clear idea of the progress they are making.

Multiple choice revision tests

These tests are provided so that students can be quickly assessed on relevant content from across multiple chapters. It is recommended to do the tests at these points:

- Test 1 after Chapter 3
- Test 2 after Chapter 6
- Test 3 after Chapter 9
- Test 4 after Chapter 11.

Examination-style papers

There are two examination-style papers, one non-calculator and one calculator, that can be used either for revision or as a mock exam. They have been set to resemble a real examination paper and can be copied for students to complete on the paper, as they would do in the real exam.

Answers and mark schemes

Comprehensive answers to the extension worksheets and multiple choice revision tests, as well as mark schemes for the examination-style papers, are included at the end of this handbook.

kerboodle

Kerboodle provides everything you need to plan, teach and track progress, including digital books, resources, assessment and automated Next Steps for each student.

We hope this handbook serves as a valuable resource, empowering you to inspire a love for mathematics and foster a deep understanding of the subject among your students.

Ian Bettison

Mathew Taylor

Content overview

The Content Overview is designed to show you what topics and content are covered in each chapter. In some cases, the content splits into two or more discrete elements and is therefore covered in more than one chapter.

Number

Chapter in the Student Book	Topic
Chapter 1	Work with different types of numbers, e.g. natural numbers, integers, primes, squares, cubes, common factors, common multiples, rational numbers, irrational numbers, reciprocals, powers and roots.
Chapter 1	Calculate squares, square roots, cubes and cube roots and other powers and roots of numbers.
Chapter 1	Calculate with and convert between the following, including in contexts: proper fractions, improper fractions, mixed numbers, decimals and percentages.
Chapter 1	Order quantities and understand the symbols $=$, $\neq$, $>$, $<$, $\geqslant$, $\leqslant$.
Chapter 1	Calculate with integers, decimals and fractions including the correct order of operations.
Chapter 1	Understand and interpret positive, zero and negative indices including using the rules of indices.
Chapter 1	Understand and calculate with numbers in standard form $A \times 10^n$ where n is a positive or negative integer, and $1 \leqslant A < 10$.
Chapter 4	Round to a given number of decimal places or significant figures and use rounded values to estimate calculations, including choosing an appropriate degree of accuracy when a question is in context.
Chapter 4	Understand and use bounds of numbers.
Chapter 4	Effective calculator use including entering values correctly and interpreting values correctly, for example, for time and money.
Chapter 6	Calculate with and convert between the following, including in contexts: proper fractions, improper fractions, mixed numbers, decimals and percentages.
Chapter 6	Simplify a ratio, divide a quantity by a ratio. Use ratios in context.
Chapter 6	Calculate a percentage of a quantity. Write one quantity as a percentage of another quantity. Solve various problems involving percentages including simple interest, compound interest, percentage increase and percentage decrease.
Chapter 6	Calculate with money including converting between currencies.
Chapter 7	Understand and use measures of rate such as pressure and density, including solving problems involving average speed.
Chapter 7	Calculate with time including in the 24-hour and 12-hour clock. Understand timetables and be able to read clocks.
Chapter 9	Understand and use Venn diagrams and set notation.

Algebra and graphs

Chapter in the Student Book	Topic
Chapter 2	Substitute into expressions and formulae.
Chapter 2	Simplify expressions, expand brackets and factorise.
Chapter 2	Understand and use rules of indices including positive, zero and negative.
Chapter 2	Construct and solve linear equations including those where x appears in the denominator as part of a linear expression. Solve simultaneous equations.
Chapter 2	Understand and use linear inequalities including representing on a number line.
Chapter 2	Understand sequences including continuing a sequence, recognising patterns, term-to-term rules, finding and using nth terms and relationships between sequences.
Chapter 6	Use, understand and draw real-life graphs such as travel graphs and conversion graphs.
Chapter 8	Use, understand and draw linear, quadratic and reciprocal graphs and solve equations graphically, for example, finding the intersection of a line and a curve.
Chapter 8	Recognise, sketch and interpret graphs of linear and quadratic functions.
Chapter 8	Use, understand and draw real-life graphs such as travel graphs and conversion graphs.
Chapter 10	Construct and solve linear equations including those where x appears in the denominator as part of a linear expression. Solve simultaneous equations.
Chapter 10	Substitute into expressions and formulae.

Coordinate geometry

Chapter in the Student Book	Topic
Chapter 3	Understand and use Cartesian coordinates.
Chapter 8	Draw graphs of linear equations.
Chapter 8	Find, use and interpret the gradient from a straight-line graph.
Chapter 8	Find, use and interpret the equation of a straight-line graph.
Chapter 8	Find, use and interpret the gradient and equation of a parallel line.
Chapter 8	Understand and use Cartesian coordinates.

Geometry

Chapter in the Student Book	Topic
Chapter 3	Use and interpret geometrical terms and vocabulary.
Chapter 3	Understand and use geometrical drawings and constructions including triangles and nets.
Chapter 3	Understand and draw scale drawings including interpreting three-figure bearings.
Chapter 3	Recognise symmetry in 2D including line symmetry and order of rotational symmetry.
Chapter 3	Use and understand geometrical properties including: • angles at a point • angles on a straight line • vertically opposite angles • sum of angles in a triangle • sum of angles in a quadrilateral • angles in parallel lines (corresponding, alternative, co-interior) • angle properties of regular polygons.
Chapter 3	Understand and use geometrical properties of circles including: • angle between tangent and radius • angle in a semi-circle.
Chapter 7	Use and interpret geometrical terms and vocabulary.
Chapter 11	Understand similar shapes including finding missing lengths.

Mensuration

Chapter in the Student Book	Topic
Chapter 7	Use and convert between units of mass, length, area, volume and capacity including in solving problems.
Chapter 7	Solve problems involving area and perimeter of triangles, rectangles, parallelograms and trapeziums.
Chapter 7	Solve problems involving circumference and area of circles, arc length and sector area.
Chapter 7	Solve problems involving surface area and volume of cuboids, prisms, cylinders, spheres, pyramids and cones.
Chapter 7	Solve problems involving area and perimeter of compound shapes, including compound solids.

Trigonometry

Chapter in the Student Book	Topic
Chapter 11	Calculate unknown sides of a right-angled triangle using Pythagoras' theorem, including solving problems in two dimensions.
Chapter 11	Calculate unknown sides and angles of a right-angled triangle using trigonometry, including solving problems in two dimensions.

Transformations and vectors

Chapter in the Student Book	Topic
Chapter 11	Draw and describe reflections, rotations, enlargements and translations.
Chapter 11	Use vector notation to describe a translation.

Probability

Chapter in the Student Book	Topic
Chapter 9	Understand and use probability including the language, notation and the probability scale from 0 to 1. Calculate the probability of an event and understand the probability of an event not happening is: (1 – the probability of the event happening).
Chapter 9	Understand and use relative frequency to estimate probability and find expected frequencies.
Chapter 9	Find the probability of combined events using sample space diagrams, Venn diagrams and tree diagrams.

Statistics

Chapter in the Student Book	Topic
Chapter 5	Use tables for statistical data.
Chapter 5	Read, interpret, draw inferences from and compare data using tables, graphs and statistical diagrams and measures, including an appreciation of limits on any conclusion made.
Chapter 5	Find the mean, median, mode and range and choose the most appropriate statistical measures.
Chapter 5	Draw and interpret the following statistical diagrams: • bar charts • pie charts • pictograms • stem-and-leaf diagrams • simple frequency distributions.
Chapter 5	Draw and interpret scatter diagrams (including understanding correlation and drawing a line of best fit).

1 Number 1

The purpose of this first chapter is to make sure that students are familiar with numbers of various kinds, so that when they appear in other topics, they will know how to use them.

The number types included in this chapter are:

- positive integers
- negative integers
- decimals
- proper and improper fractions
- powers of 10
- rational and irrational numbers
- numbers in standard form.

Students will also study these other basic skills:

- reading scales
- using the correct order of operations
- working with indices
- finding the lowest common multiple and highest common factor
- finding reciprocals

The emphasis in this chapter is on working without a calculator.

After studying this chapter, students should be ready to use these numbers and skills in other contexts.

Prerequisite knowledge

There is very little prerequisite knowledge required for studying this chapter.

Students should be able to count, know their basic multiplication tables and be familiar with place value.

Investigations, activities and puzzles

The biggest number

A calculator has the following buttons:

+	−	×	÷	(	)	y^x	=

The only digits buttons that work are the '1', '2' and '3'.

1. You can press any button, but only once.
   What is the biggest number you can get?
2. Now the '1', '2', '3' and '4' buttons are working.
   What is the biggest number you can get?
3. Investigate what happens as you increase the number of digits you can use.

Notes on this activity
This activity is a bit of fun that gives students an opportunity to experiment with the buttons on their calculators.

Extension

There are many ways in which this activity could be extended.
For example:

- give the students a target number to try to make
- allow more functions on their calculators to be used

Happy numbers

Take a number.

Square each digit and add the answers together.

Repeat with your new number. Keep going until you reach 1, or a number that you have reached before.

For example, for 23:

$$2 \quad 3$$
$$2^2 + 3^2 = 1 \quad 3$$
$$1^2 + 3^2 = 1 \quad 0$$
$$1^2 + 0^2 = 1$$

Because the sequence ends at 1, mathematicians call 23 a 'happy' number.

Investigate for other numbers.

Here are a few suggestions: 70, 85, 49, 44, 14, 15, 94.

Which of these are happy?

Notes on this activity
There are many different types of numbers that mathematicians have given names to. Happy numbers are just one example.

Extension

You could also extend this activity in the following ways:

- There are seven prime numbers between 1 and 100 that are happy. Ask students to try to find them.
- Change the rule. Instead of squaring the digits, cube them.

$$2 \quad 1$$
$$2^3 + 1^3 = 0 \quad 9$$
$$0^3 + 9^3 = 7 \quad 2 \quad 9$$
$$7^3 + 2^3 + 9^3 = 1 \quad 0 \quad 8 \quad 0$$
$$1^3 + 0^3 + 8^3 + 0^3 = 5 \quad 1 \quad 3$$
$$5^3 + 1^3 + 3^3 = 153$$

  And now we are stuck because 153 leads to 153 again.

  Investigate for numbers of your own choice. Do any numbers lead to 1?
- Find out about other kinds of numbers that have been given names.

Consecutive sums

What numbers can you make by adding together consecutive positive integers?

For example, you can make all the numbers between 20 and 30 in the following ways:

$20 = 2 + 3 + 4 + 5 + 6$	$21 = 10 + 11$	$22 = 4 + 5 + 6 + 7$
$23 = 11 + 12$	$24 = 7 + 8 + 9$	$25 = 12 + 13$
$26 = 5 + 6 + 7 + 8$	$27 = 13 + 14$	
$28 = 1 + 2 + 3 + 4 + 5 + 6 + 7$	$29 = 14 + 15$	$30 = 6 + 7 + 8 + 9$

Are there any numbers that you can't make by adding consecutive positive integers?

Notes on this activity

Students will discover that the numbers that they can't make are the powers of 2.

They could then try to work out why this is.

Along the way, they will discover that all odd numbers are the sum of two consecutive integers. Then they can look for a pattern in the sums of three consecutive integers.

It is quite a structured activity that leads to some interesting results for students of all abilities.

Glossary

Cube number

A cube number is the result obtained when a number is multiplied by itself and then multiplied by itself again.

Example

8 is a cube number as $2 \times 2 \times 2 = 8$

Cube root

The cube root of a number is another number which, when cubed, will equal the first number. The cube root of 1 is 1.

Examples

The cube root of 8 is 2 since $2 \times 2 \times 2 = 8$

$\sqrt[3]{\ }$ is the symbol meaning 'the cube root of the number given'.

$\sqrt[3]{8} = 2$

$\sqrt[3]{4.096} = 1.6$ since $1.6 \times 1.6 \times 1.6 = 4.096$

Decimal fraction

A decimal fraction is a way of expressing values of fractions less than 1 using the normal decimal place-value system extended to the right of the units column to give values of tenths, hundredths, and so on.

Example

0.376 means $\frac{3}{10} + \frac{7}{100} + \frac{6}{1000} = \frac{376}{1000}$

Factor

A number that divides exactly into another number is a factor of that number.

Example

3 is a factor of 24 because 8 is a whole number. However, 5 is not a factor of 24 because 4.8 is not a whole number.

Fraction

A fraction is a measure of how something is to be divided up or shared out. There are four principal ways of expressing a fraction: common, decimal, percentage and ratio.

Examples

$\frac{1}{2}$ 0.5 50% 1:2

Highest common factor

The highest common factor of two numbers is the highest number that is a factor of both of those numbers.

Example

The factors of 60 are:
1, 2, 3, 4, 5, 6, 10, 12, 15, 20, 30, 60

The factors of 66 are:
1, 2, 3, 6, 11, 22, 33 66

The highest number in both lists is 6, so 6 is the highest common factor of 60 and 66.

Note that 60 and 66 also have other factors in common, but 6 is the highest.

Integer

A number made from the natural numbers (including 0) by putting a positive or a negative sign in front. The positive sign is often omitted.

The integers are:
..., −5, −4, −3, −2, −1, 0, 1, 2, 3, 4, 5, ...

Inverse operation

An operation that does the opposite of another operation.

Example

Subtraction is the inverse of addition.

Irrational numbers

Numbers that can only be written in numerical form (using no symbols) as a never-ending, non-repeating decimal fraction. An irrational number cannot be written in the form of a rational number.

Example

1.23456789101112131415161718192021222324...

This number is generated by writing out the natural numbers; it could go on forever without repeating. It is an irrational number.
The square root of any prime number is irrational, as is π.

Lowest common multiple

The lowest common multiple of two numbers is the lowest number that is a multiple of both of those numbers.

Examples

The multiples of 3 are:
3, 6, 9, 12, 15, 18, 21, 24, 27...

The multiples of 4 are:
4, 8, 12, 16, 20, 24, 28...

The lowest number in both lists is 12, so 12 is the lowest common multiple of 3 and 4.

Note that 3 and 4 also have other multiples in common but 12 is the lowest.

Multiple

When you repeatedly add a number to itself you produce multiples of the original number.

Example

If I start with 7 and keep adding 7, I will get:
14, 21, 28, 35...

These are called 'multiples of 7'.

The first multiple of a number is the number itself.

Natural numbers

The set of numbers 1, 2, 3, 4, 5, 6, ... as used in counting. It is a matter of choice whether 0 is included or not.

Negative numbers

Numbers less than zero.

Power or Index

Index notation is a way of indicating how a number (or symbol) must be operated on, using another number written as a superscript to the first; the superscript number is called an index. When the index is a positive whole number, that number indicates how many of the first number or symbol must be multiplied together. When the index is a unitary fraction then it indicates a root has to be found. When the index is a non-unitary fraction then it indicates a power and a root have to be found.

Examples

$A^2 = A \times A$ $\qquad$ $5^3 = 5 \times 5 \times 5 = 125$

$9^{\frac{1}{2}} = \sqrt{9} = 3$ $\qquad$ $8^{(2/3)} = 4$

Prime numbers

Numbers that have two factors, one and itself. The list continues forever. Note that 1 is not a prime number because it has only one factor.

Examples

2, 3, 5, 7, 11, 13, 17, 19, 23, 29....

Prime factor

Any prime number which is a factor of another number is called one of its prime factors.

Prime factors are important in mathematics because any positive number greater than 1 which is not a prime number can be expressed as a product of prime factors in only one way.

Examples

3 is a factor of 30. Since 3 is prime, it is one of the prime factors of 30.

30 cannot be made by multiplying together any other set of prime numbers.

Reciprocal

The reciprocal of a number is what you get if you divide 1 by that number.

If the original number is a fraction, you can find its reciprocal by turning the fraction upside down.

Examples

The reciprocal of 1 is 1.

For any whole number n, its reciprocal is $\frac{1}{n}$.

The reciprocal of $\frac{2}{3}$ is $\frac{3}{2}$.

Rational numbers

Numbers that can be written in the form $\frac{a}{b}$ where a and b are both integers and b is not zero.

Examples

-4.5 $\qquad$ $1\frac{1}{3}$

0.0909090909...

3.142857 $\qquad$ 8

Square number
A square number is the result obtained when a number is multiplied by itself.

Example
25 is a square number as $5 \times 5 = 25$

Square root
A square root of a number is another number which, when squared, will be equal to the first number. The square roots of 1 are 1 and -1.

Examples
One square root of 16 is 4 because $4 \times 4 = 16$; another is -4.

$\sqrt{}$ is the symbol meaning 'the positive square root of the number given'.

$\sqrt{49} = 7$ $(7 \times 7 = 49)$

$\sqrt{3.24} = 1.8$ $(1.8 \times 1.8 = 3.24)$

Standard form
A way of displaying a number in the form of a first number, whose value lies between 1 and 10, and a second number, which is always 10 with a suitable integer index, so that the two numbers multiplied together are equal to the value of the intended number.

Examples
$1.436 \times 10^2 = 143.6$

$1.436 \times 10^{-3} = 0.001\,436$

Extension worksheet

This worksheet develops the work on prime numbers, highest common factors, and lowest common multiples.

1. Write down a list of all of the prime numbers under 50.

An efficient way of testing whether a number is prime or not is to try to divide it by all the primes that are smaller than the square root of the number. If none divide into it, the number is itself prime. Can you explain why this works?

For example:

157

$\sqrt{157} = 12.53$ (to 2 decimal places)

157 cannot be divided by 2, 3, 5, 7 or 11 so it is prime.

2. Which of the following numbers are prime?

73   91   231   1377   2499

Prime factor trees and Venn diagrams can be used to find the lowest common multiple and highest common factor of two numbers.

For example:
$24 = 2^3 \times 3$ and $30 = 2 \times 3 \times 5$

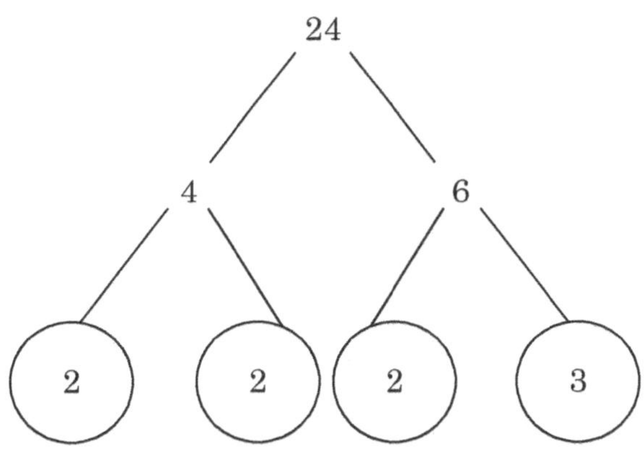

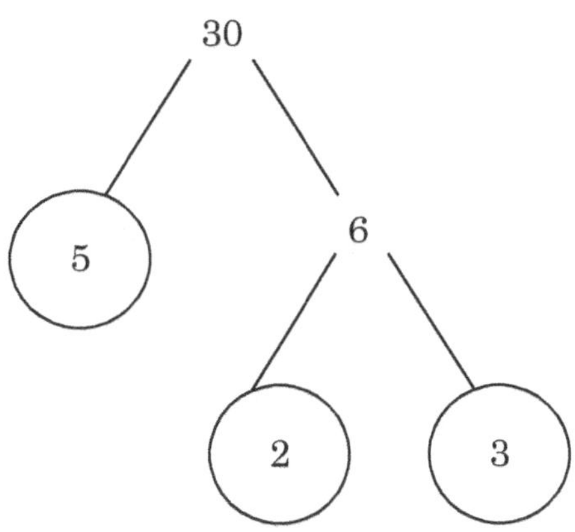

Write the *common* prime factors in the middle part of the Venn diagram and the other factors in the other parts of the circles:

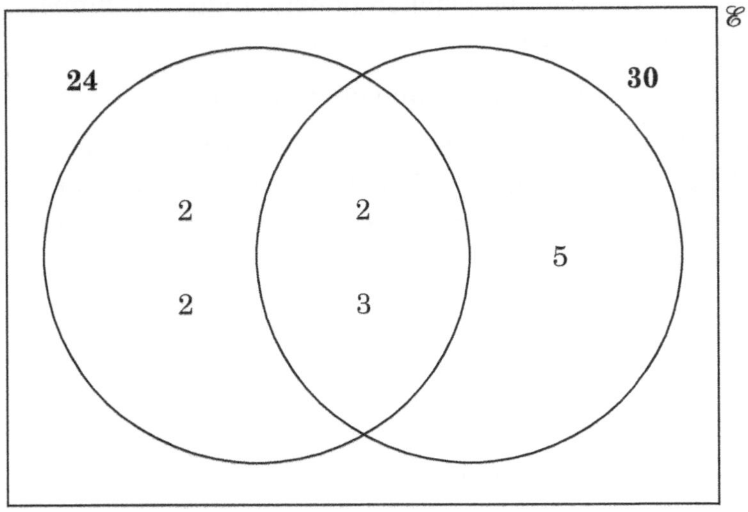

The HCF is the product of the middle numbers ($2 \times 3 = 6$) and the LCM is the product of all the numbers ($2 \times 2 \times 2 \times 3 \times 5 = 120$).

3. Use this method to find **(i)** the HCF and **(ii)** the LCM for the following pairs of numbers:

a) 40 and 60

b) 25 and 110

c) 120 and 150

d) 64 and 100

e) 250 and 320

f) 125 and 300

(Answers on page 118)

2 Algebra 1

This chapter introduces students to the concept of algebra, starting by teaching them the important vocabulary, such as the words 'term', 'expression', 'coefficient' and 'constant'.

The concept of 'like terms' is considered, and then students practise substituting values into expressions.

The concept of a term is then looked at in the context of linear sequences and finding an nth term rule. Sequences from patterns are also considered.

The chapter then returns to algebraic techniques, which include:

- expanding brackets
- factorising
- solving equations with x on one side
- solving equations with x on both sides
- solving worded problems using algebra.

After briefly considering indices, the chapter then ends by looking at linear inequalities.

Prerequisite knowledge

Before attempting this chapter, students should be:

- familiar with the rules of BIDMAS
- able to work confidently with negative numbers
- able to work with fractions.

Investigations, activities and puzzles

Scales

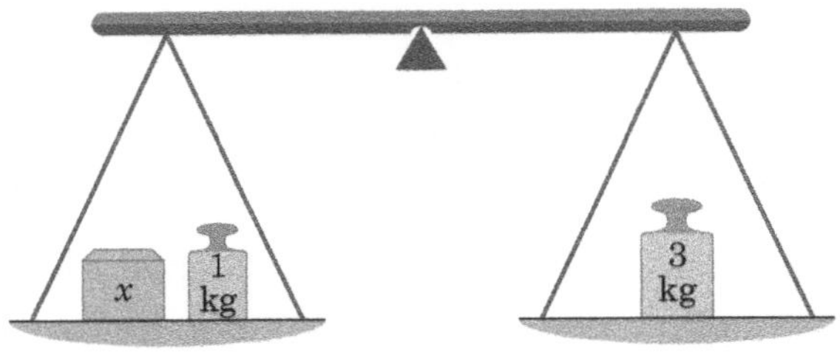

In the diagram the mass of the package x is measured using two masses.

If the scales are balanced, x must be 2 kg.

1. How can you measure all the masses from 1 kg to 10 kg using three masses: 1 kg, 3 kg, 6 kg?
2. It is possible to measure all the masses from 1 kg to 13 kg using a different set of three masses.
   What are the three masses?
3. It is possible to measure all the masses from 1 kg to 40 kg using four masses.
   What are the masses?

Notes on this activity

In terms of algebra content, the main ideas in this activity are that x stands for a number and that this number can change.

It would be a good idea to discuss the first question in this activity with the students.

How would they actually do this for a randomly chosen value of x within the range being considered?

Once Question **1** has been answered by the class, they can move on to Question **2**, which focuses on how the three values for the masses are chosen. How did they choose them? Was it simply trial and improvement or do they have any strategies?

Question 3 extends the problem so that they are using more masses to measure a greater number of kilograms.

Extension

There are many other questions that could be considered about this problem. If we are concerned with measuring all the masses from 1 kg to n kg, here are some of the questions that students could consider:

- What is the greatest number of kilograms that can be measured using three masses? Why does this limit exist?
- What is the greatest number of kilograms that can be measured using four masses? (an extension of the previous question)
- At what point would you need to introduce a fifth mass?
- How do you choose which masses to use?
- Are there some combinations of masses that are particularly bad choices?

An expanding diagram

Look at the series of diagrams below.

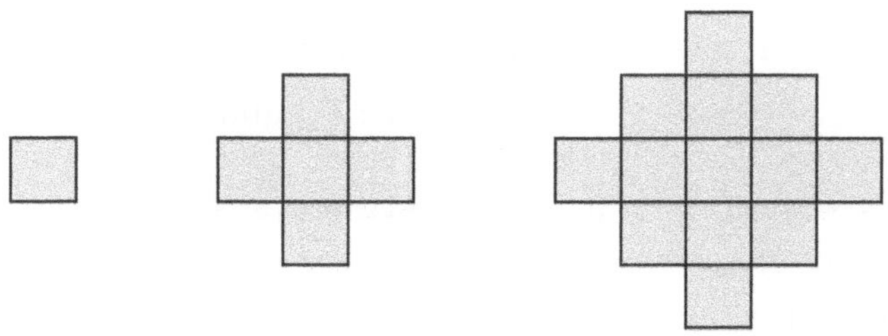

1 2 3

Each new diagram is obtained by drawing squares all around the outside of the previous diagram. For each diagram, count the number of squares it contains.

Can you predict the number of squares in the next diagram?

What about the diagram after that?

Can you find a rule that will enable you to predict the number of squares for any member of the sequence of diagrams.

Can you predict the number of squares in:

a) the 12th diagram

b) the 20th diagram?

Notes on this activity

One way to make sense of these diagrams is to notice that the nth diagram contains n diagonal rows of n squares (the black squares) plus $(n - 1)$ diagonal rows of $(n - 1)$ squares (the remaining squares).

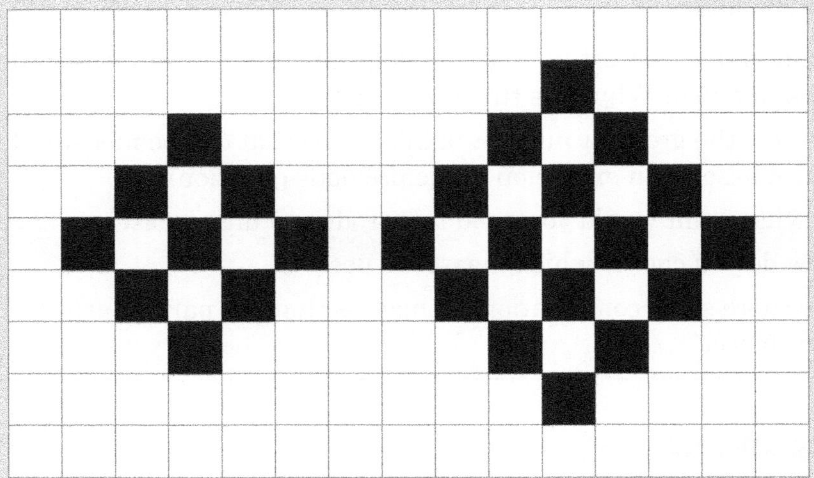

This means that the nth diagram will contain

$$\begin{aligned} n^2 + (n-1)^2 &= n^2 + n^2 - 2n + 1 \\ &= 2n^2 - 2n + 1 \text{ squares.} \end{aligned}$$

Although this sequence is quadratic, students should be able to derive it directly from the diagrams, rather than having to analyse the quadratic sequence algebraically.

Glossary

Equation

An equation is a statement that two expressions (one of which may be a constant) have the same value.

Examples

$2x + 7 = 15$ and $3(x + 5) = 3x + 15$
are both equations.

Expression

An expression in algebra is most often a collection of quantities, made up of constants and variables, linked by signs for operations and usually not including an equals sign.
In practice, it is an imprecise word and is used very loosely.

Examples

$x + y$ $\quad$ $3 + x^2 - y$
$4(x - y)$ $\quad$ $3x^2 + 5y$
are all expressions.

Inequality

Inequalities are like equations except they tell you when something is less than, less than or equal to, greater than, or greater than or equal to something else.

Examples

$x > 3$ means that x is greater than 3.

This inequality defines a range of possible values for x, rather than a single value.
In this inequality, 3 is a lower bound for the value of x.

Inequalities can also define a range with a lower bound and an upper bound.

$1 < x < 10$ tells you that x can be any number between, but not including, 1 and 10.

Sequence

A sequence is a set of numbers or objects made and written in order according to some mathematical rule.

Example

1, 4, 9, 16, …

Substitution

A substitution in algebra is done by replacing one expression, or part of an expression, by something of equivalent value so that the overall truth of the original expression is unchanged.

Example

Given $4y + 3x = 22$ and $y = 2x$,
then the second expression can be substituted for y in the first to give $4(2x) + 3x = 22$, which simplifies to $11x = 22$

Term

The terms in a simple algebraic expression are the quantities that are linked to each other by means of + or – signs. In more complicated expressions the word 'term' is given a much looser meaning.

Example

$5x^2 + 3x - xy + 7$ has four terms.

Extension worksheet

This worksheet develops the work on sequences to consider non-linear sequences and their generating functions.

Find the next two terms in the following sequences and describe the rule you have used in words.

1. 1, 4, 9, 16, ...

2. 2, 5, 10, 17, ...

3. 1, 3, 6, 10, ...

4. 3, 5, 8, 12, ...

5. 10, 20, 32, 46, ...

6. 1, 2, 4, 8, ...

7. 1, 3, 9, 27, ...

8. 3, 6, 12, 24, ...

The questions below give formulae for generating the terms of a number of sequences. Write out the first five terms of each sequence.
Using 'T_n =' is a way of saying 'the terms of a sequence are generated by...' and then the formula follows. To generate the first five terms, substitute values for n of 1 to 5 into the right-hand side of each formula.

9. $T_n = n^2 - 1$

10. $T_n = 2^n$

11. $T_n = n^2 + 3$

12. $T_n = n^2 + n$

13. $T_n = 4^{n-1}$

14. $T_n = n^3$

15. $T_n = 2^n - 3$

16. $T_n = n^2 + 3n - 1$

17. $T_n = 2n^2 - 4n + 1$

18. $T_n = 3^{n+2}$

19. Investigate the:

a) Fibonacci sequence

b) tetrahedral number sequence

c) pentagonal number sequence.

(Answers on page 118)

3 Shape and space 1

Having looked at number and algebra, this chapter now moves the focus to geometry.

It begins by checking that students can draw basic shapes by accurately measuring angles and lines.

Three-dimensional shapes are then briefly considered, along with the idea of a net.

Angles are then studied in detail, making sure that students know:

- the names of different kinds of angles
- simple angle sums
- how different angles are related in situations involving parallel lines
- how to find angles in larger polygons
- how to calculate interior and exterior angles
- that the angle in a semi-circle is a right angle.

Along the way, quadrilaterals are looked at in more detail, focusing on the various properties of different shapes.

While looking briefly at circles, various circle vocabulary words are introduced.

Students then consider the concepts of reflection and rotational symmetry, before completing an exercise involving scale drawings.

This then leads into the final part of the chapter, which is on bearings.

Prerequisite knowledge

Before attempting this chapter, students should:

- be able to use a protractor and measure accurately with a ruler
- be familiar with the names of basic polygons
- understand what is meant by an angle.

Investigations, activities and puzzles

Constructing shapes

Construct an equilateral triangle of side length 7 cm in the following way, using only a pencil, a straight edge, and a pair of compasses.

1. Draw a line segment AB of length 7 cm.
2. Set your compasses to 7 cm, and then draw an arc from point A.
3. Keeping your compasses at 7 cm, draw another arc from B that intersects the first arc.
4. Let C be the point where the two arcs intersect. Join AC and BC to complete the triangle.

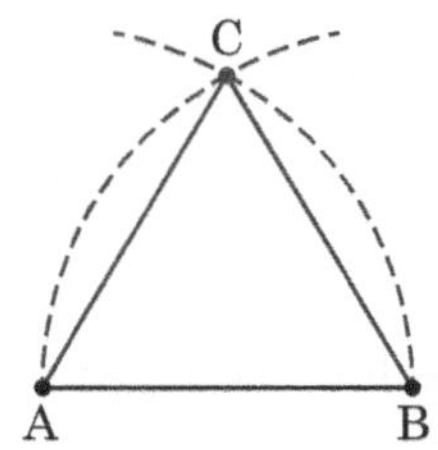

What other shapes can you construct?

Notes on this activity

This is a great activity. As well as making students more confident when using compasses, the idea of being able to accurately construct quite complex shapes such as pentagons and decagons is something that a lot of them really enjoy.

Look online and find out which shapes is it possible to construct using only a pencil, a straight edge and a pair of compasses.

There are a lot of really good tutorial videos that you can show to students, in which the entire construction is demonstrated.

Extension

Since this is an exercise in geometry, there are many ways in which it can be extended, such as:

- considering which shapes can be constructed accurately and which ones cannot
- following instructions for how to construct good approximations of the ones that cannot be constructed accurately (e.g. heptagon and nonagon).

Nets

Both of these nets can be folded to make a cube.

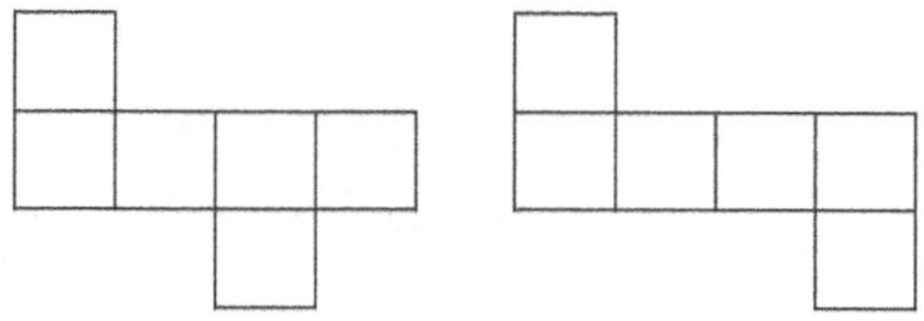

How many different nets does a cube have?

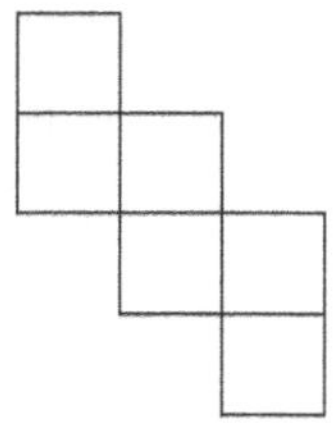

Is this a net for a cube?

Notes on this activity

This is an interesting little activity. You should tell students that a rotated version of one net does not count as a new net.

Maps

Bearings are important for ships navigating at sea or planes flying in the air.

A map projection is a method for transforming the surface of a sphere into a flat surface. Over the centuries, different map projections have been created to help people find their way around.

The most popular one is Mercator's projection from 1569. It became popular because although some things are distorted, all the bearings that you would need to travel on to get from one place to another are kept accurate.

Investigate other map projections. What are their strengths and weaknesses?

Glossary

Angle
An angle is made when two straight lines cross or meet each other at a point. Its size is measured by the amount one line has been turned in relation to the other.

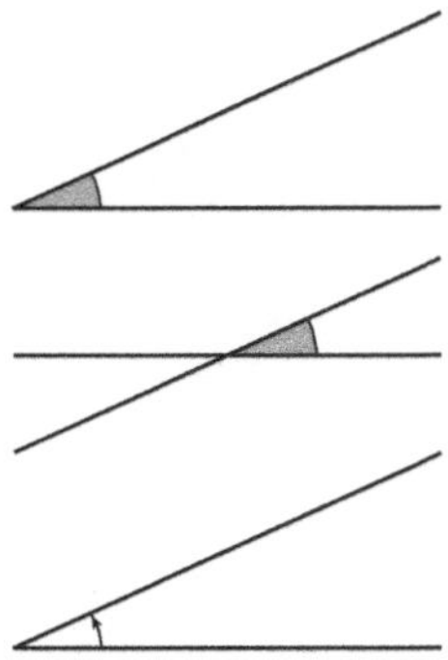

Angle Sum
The angle sum of a polygon is the total of all its interior angles added together.

angle sum of any polygon = (180° × number of sides) – 360°

Examples
For a triangle the angle sum is 180°;
for a quadrilateral it is 360°.

Bearing
The bearing of position B from position A is the direction measured in a clockwise direction from north in which someone travelling in a straight line from A to B must go. *Care should be taken when using 'from' and 'to' when working with bearings.*

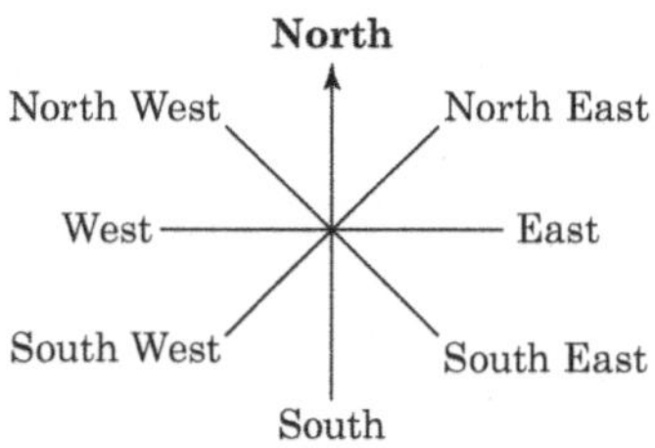

The 8 points of the compass by name

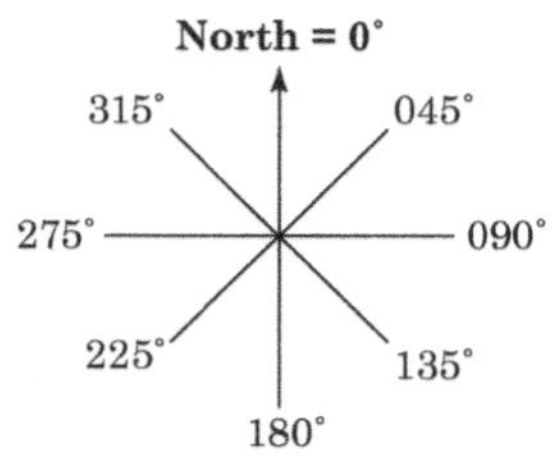

The 8 points of the compass measured in degrees clockwise from north

Chord
A chord of a circle is any straight line drawn across the circle, beginning and ending on the curve making the circle. A chord which passes through the centre is also a diameter.

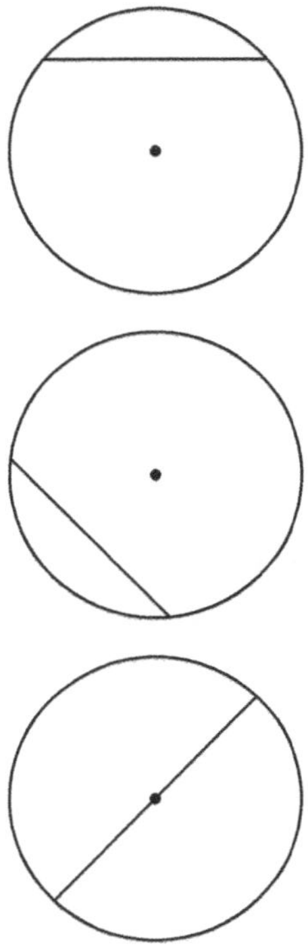

Circumference
The circumference of a circle is the distance measured around the curve which makes the circle.

circumference = π × diameter = πd

Congruent
Geometrical figures are said to be congruent if they are the same shape and size. One shape can be fitted exactly over the other, being turned around and/or over as necessary. Such shapes are also described as being identical or equal.

Example
These four triangles are congruent:

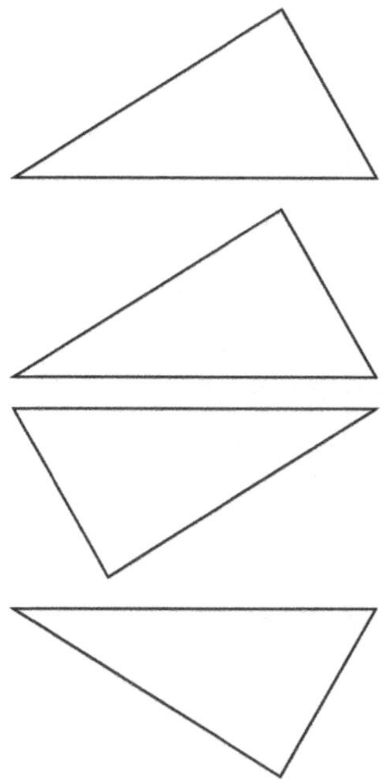

Diameter
A diameter of a circle is a straight line passing through the centre and which touches the curve forming the circle at each of its ends.

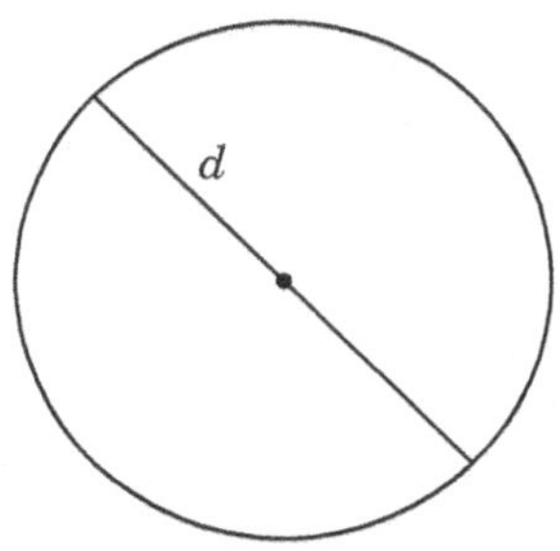

Exterior Angle
The exterior angle of a polygon is the angle formed outside the polygon between any one side and the side adjacent to it, extended.

size of exterior angle = 180° − interior angle

Factor
A factor is a number which divides exactly into another number. 1 is a factor of every number and every number is a factor of itself.

Examples
1 is a factor of 5;
3 is a factor of 6;
4 is a factor of 12;
7 is a factor of 7;
2 and 17 are factors of 68

Highest Common Factor (HCF)
The highest common factor of two (or more) numbers is the common factor of all those numbers which has the greatest value.
The HCF may be one of the actual numbers.

Examples
The HCF of 12 and 18 is 6
The HCF of 12 and 17 is 1
The HCF of 5, 15 and 30 is 5

Interior Angles
An interior angle is the angle formed inside a polygon between two adjacent sides (i.e. sides that are next to each other).

Kite
A kite is a quadrilateral which has two pairs of adjacent sides (i.e. sides that are next to each other) of the same length, with all interior angles less than 180°. It has one line of symmetry and its diagonals cross each other at right angles.

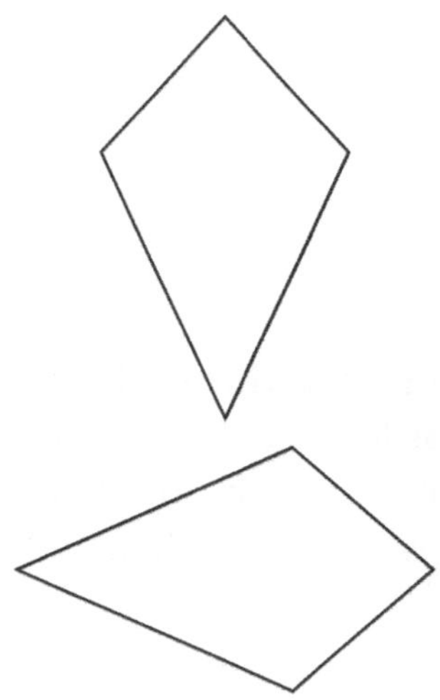

Line Symmetry
Line symmetry is the symmetry of a plane shape (i.e. a flat or 2D shape) which can be folded along a line so that one half of the shape fits exactly on the other half. A shape can have several lines of symmetry.

Examples of shapes with line symmetry

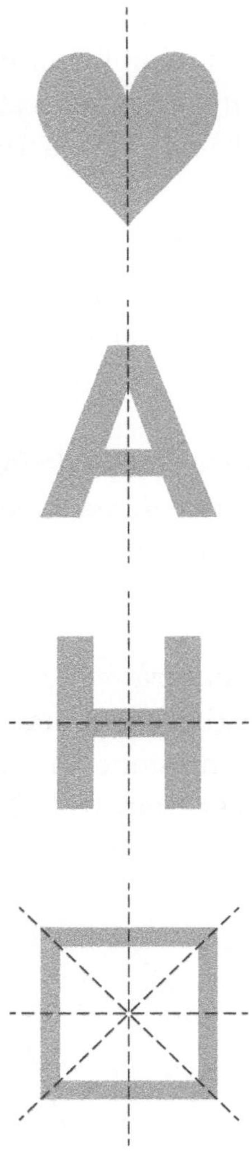

Lowest Common Multiple (LCM)

The lowest (or least) common multiple of two (or more) numbers is the smallest possible number into which all of them will divide exactly.

Example

The LCM of 3, 4 and 8 is 24

Multiple

A multiple is a number made by multiplying together two other numbers. *A number is a multiple of any of its factors.*

Example

12 is a multiple of 2 since $12 = 2 \times 6$

Net

A net is a flat 2D shape (for example, an arrangement of polygons connected at their sides), which can be folded up to make a 3D shape (for example, a polyhedron).
For a polyhedron, there is always more than one way of doing this.

Parallelogram

A parallelogram is a quadrilateral (i.e. a four-sided shape) which has two pairs of parallel edges. It has rotational symmetry of order 2, and its diagonals bisect each other. Usually one pair of edges is longer than the other pair, no interior angle (i.e. corner) is a right angle and it has no lines of symmetry.

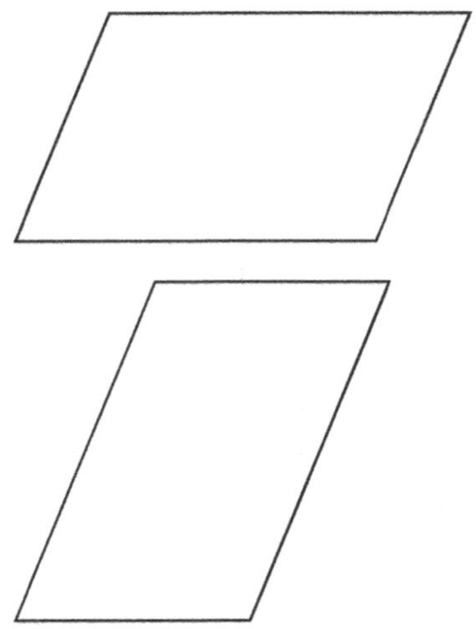

Polygon

A polygon is a plane (i.e. flat) shape completely enclosed by three or more straight sides. The sides do not cross one another, and the word is not often used for shapes having fewer than five sides. Polygons are named by the number of sides or angles they have – see table on next page. A regular polygon has all sides of equal length and all angles of equal size.

Example

A regular pentagon

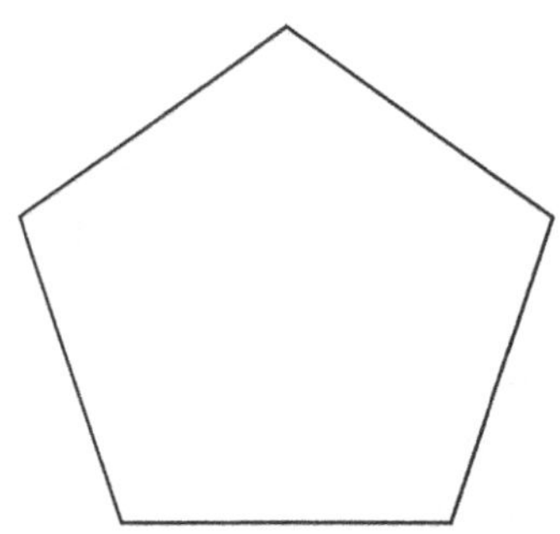

Data for regular polygons		
Number of sides	**Name**	**Interior angle**
3	triangle	60°
4	quadrilateral	90°
5	pentagon	108°
6	hexagon	120°
7	heptagon	128.57°
8	octagon	135°
9	nonagon	140°
10	decagon	144°
11	undecagon	147.27°
12	dodecagon	150°

Radius

The radius of a circle is the distance from the centre to the curve which makes the circle.
A radius is any straight line from the centre to the curve.

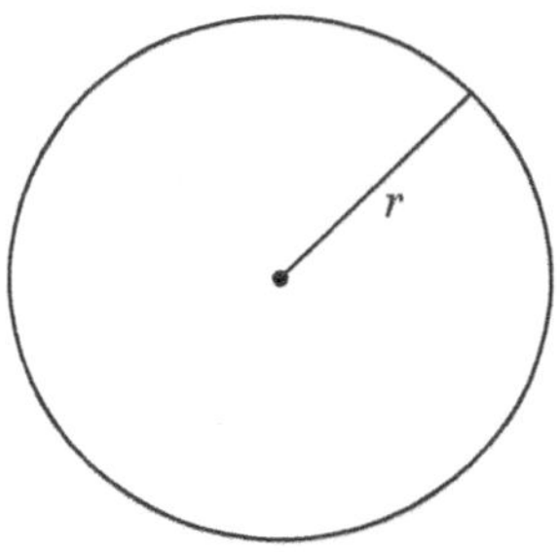

Rectangle

A rectangle is a quadrilateral (i.e. four-sided shape) in which every interior angle (i.e. corner) is a right angle.

Rhombus

A rhombus is a quadrilateral whose sides are all the same length. Its diagonals bisect each other at right angles and both are also lines of symmetry. Usually, the interior angles are not right angles. It is sometimes referred to (incorrectly) as a diamond, lozenge, or rhomb.

Rotational symmetry

Rotational symmetry is the symmetry of a shape that may be turned and fitted on to itself somewhere other than in its original position.

Examples with centres about which the shape is turned

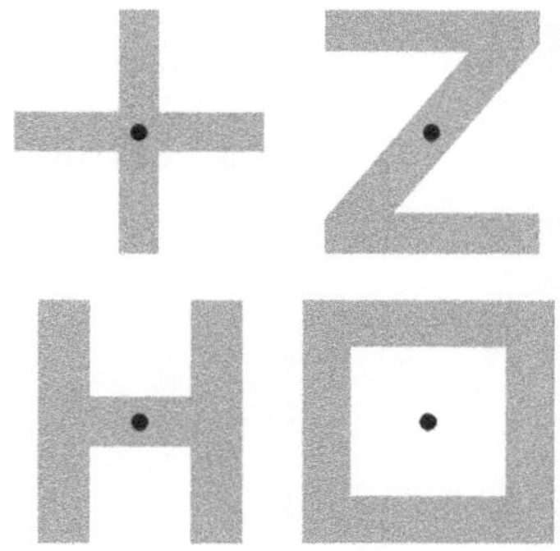

Scale

When making a drawing of an object that is meant to be in proportion to the size of the object, and from which measurements can be taken, a scale is used to fix the ratio between the actual measurements on the object and those in the drawing.

Trapezium

A trapezium is a quadrilateral (i.e. four-sided shape) with only one pair of parallel sides.

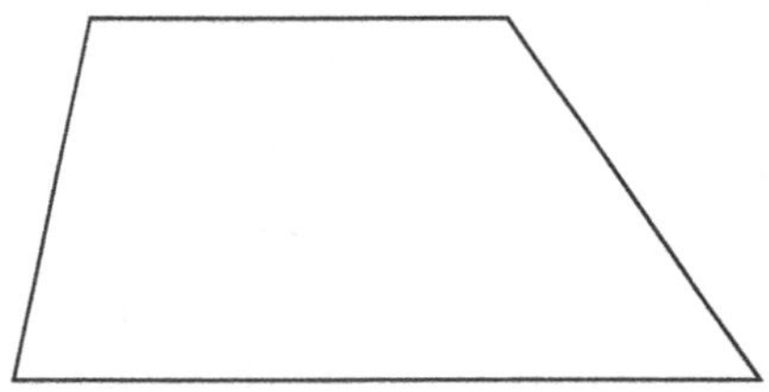

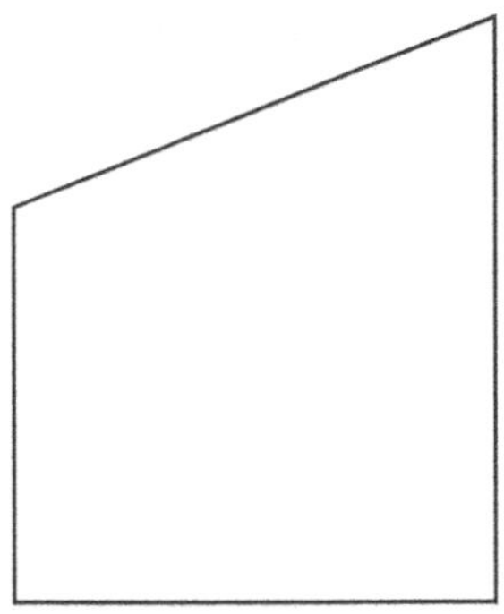

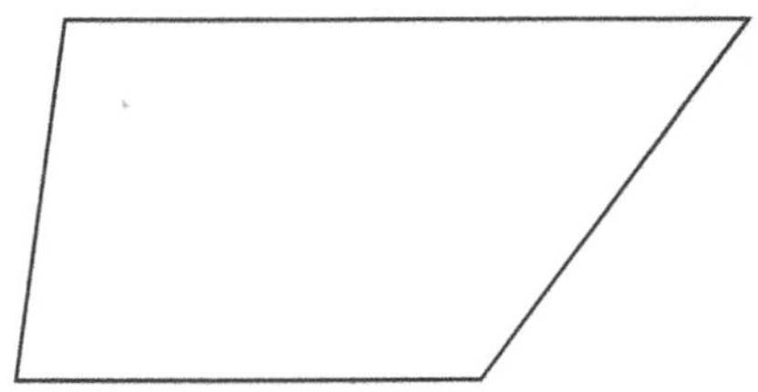

Vertex

A vertex is a point where two sides of a polygon meet to form a corner.

Extension worksheet

This worksheet develops the work on rotational symmetry to consider different orders of rotation.

1. What is the least number of extra squares you need to shade to make this pattern have rotational symmetry of order 2?

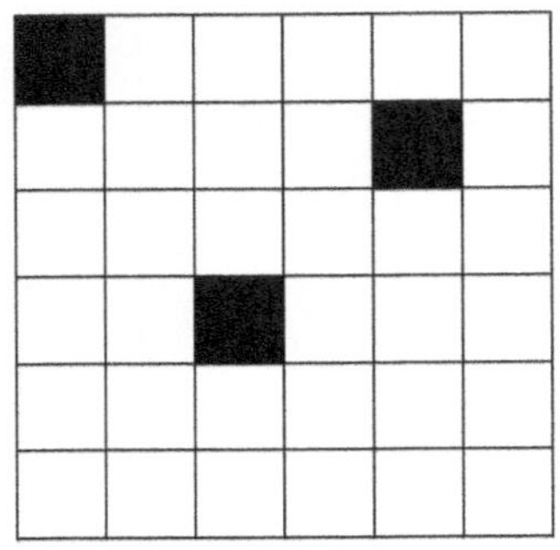

How many squares would you need to shade to make it have rotational symmetry of order 4?

What is the maximum number of squares you could shade for it to have only rotational symmetry of order 2 and not of order 4?

2. Consider a smaller, 3 by 3, square. How many ways are there of shading in squares so that it has rotational symmetry of order 2 but not of order 4?

Use the grids on the next page to work this out. Two examples have been done for you.

Can you predict how many ways you can shade a 4 by 4 square so that it has rotational symmetry of order 2 but not of order 4?
Is there a connection?

What would happen with bigger and bigger grids?

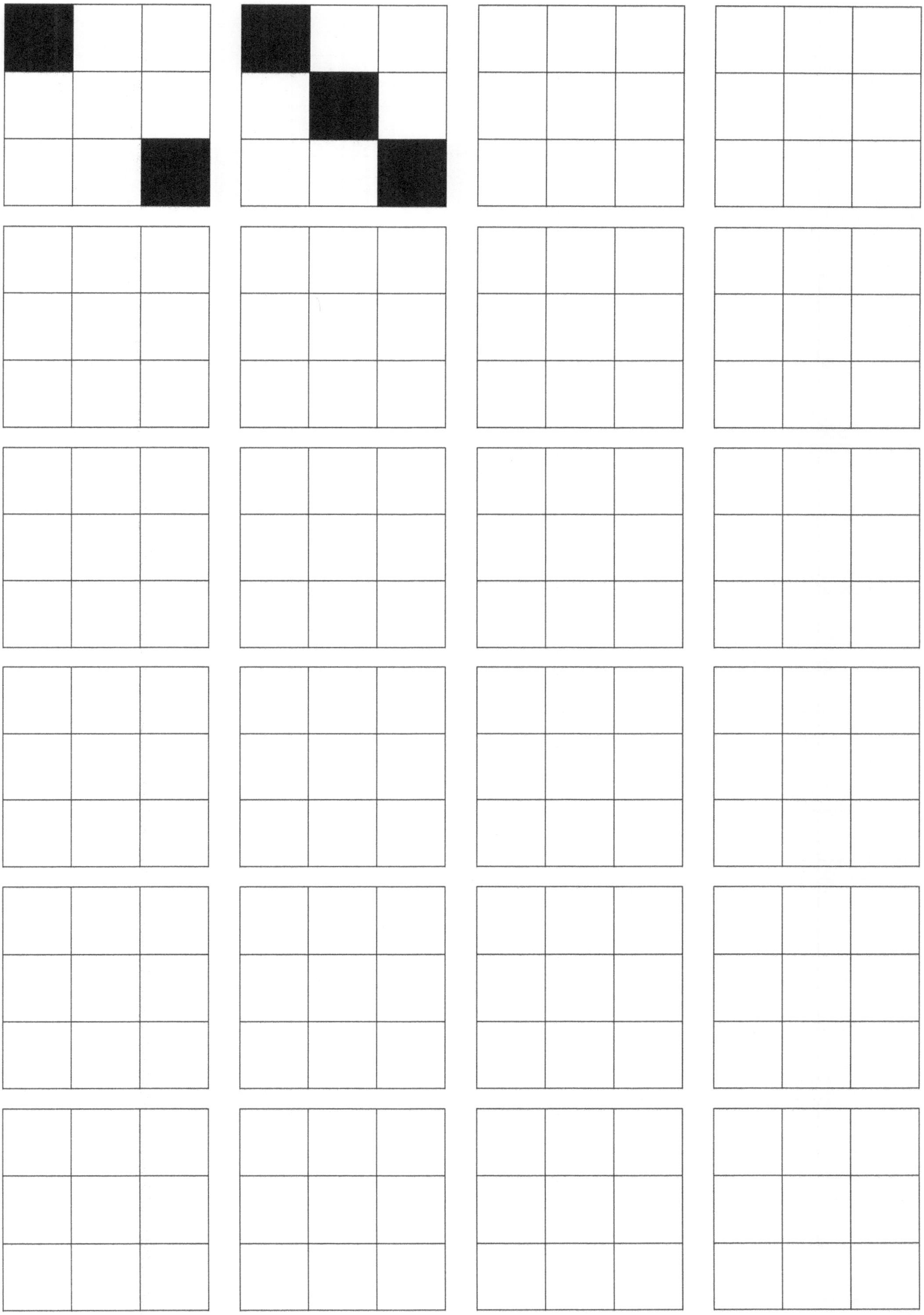

(Answers on page 119)

4 Number 2

This is a short chapter that focuses mostly on approximating values.

Students learn to approximate in the following ways:

- rounding numbers
  - using significant figures
  - using decimal places
- estimating calculations by using rounded numbers.

They then learn how to interpret approximate values by considering the upper and lower bounds of what those values could actually represent.

The chapter ends with an exercise designed to make sure that students can accurately use their calculators to perform tricky calculations. Some questions on time and money are provided so that students can use their calculators to perform some calculations that have a real-world context.

Prerequisite knowledge

Before attempting this chapter, students should:

- be confident with decimal numbers and place value
- be familiar with metric units.

Investigations, activities and puzzles

Making money

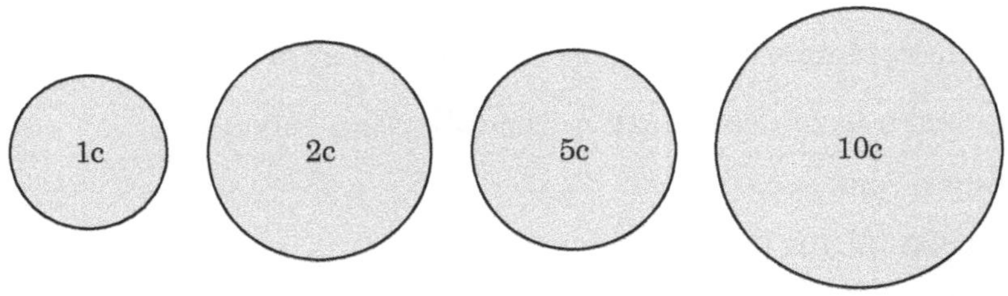

You can combine these coins to make many different totals.

How many different total values can you make? You do not need to use all the coins each time, but you can use each coin only once.

Suppose you now have:
one 1c, one 2c, one 5c, one 10c, one 20c, one 50c and one $1 note.

How many different totals can you make now?

Notes on this activity

This is an interesting activity with plenty of scope for expansion.

If you want the totals to be all the values between 1c and 99c, are the coins listed above the most efficient? Would there be a better set of values to have?

If you are allowed as many coins as you like of the given denominations, which total between 1c and 99c requires you to use the most coins to make it?

What is the smallest set of coins you would need for the totals to be all the values between 1c and 99c? What would those coins be?

Calculator words

On a calculator, work out $9508^2 + 192^2 + 10^2 + 6$.
If you turn the calculator upside down and use a little imagination, you can see the word 'HEDGEHOG'.

Find the words given by the clues below.

1. $19 \times 20 \times 14 - 2.66$ (not an honest man)
2. $(84 + 17) \times 5$ (dotty message)
3. $904^2 + 89\,621\,818$ (prickly creature)
4. $(559 \times 6) + (21 \times 55)$ (what a surprise!)
5. $566 \times 711 - 23\,617$ (eat it quickly)
6. $\dfrac{9999 + 319}{8.47 + 2.53}$ (sit up and plead)

7. $\frac{2601 \times 6}{4^2 + 1^2}$; $(2 + 3) \times 11 \times 10^2 + 9 - 1$ (two words for the person in charge)

8. $7 \times 10 \times 100 + 2 \times 9 \times 41$ (you can ring it)

9. $\frac{27 \times 2000 - 2}{0.63 \div 0.09}$ (not quite a mountain)

10. $(5^2 - 1^2)^4 - 14\,239$ (a name)

11. $48^4 + 102^2 - 4^2$ (pursuits)

12. $615^2 + (7 \times 242)$ (a small laugh)

13. $(130 \times 135) + (23 \times 3 \times 11 \times 23)$ (wobbly)

14. $164 \times 166^2 + 734$ (almost big)

15. $8794^2 + 25 \times 342.28 + 120 \times 25$ (thin skin)

16. $0.08 - (3^2 \div 10^4)$ (ice house)

17. $235^2 - (4 \times 36.5)$ (shiny surface)

18. $3 \times 17 \times (329^2 + 2 \times 173)$ (unable to stand)

19. $230^2 + \frac{1}{2} \times 230 + 30$ (these go on feet)

20. $(23 \times 24 \times 25 \times 26) + (3 \times 11 \times 10^3) - 20$ (help)

21. $(22^2 + 29.4) \times 10$; $(3.03^2 - 0.02^2) \times 100^2$ (four words, Goliath)

22. $1.25 \times 0.2^6 + 0.2^2$ (don't cry)

23. $(3^3)^2 + 2^2$ (wriggler)

24. $14 + (5 \times (83^2 + 110))$ (bigger than a duck)

25. $2 \times 3 \times 53 \times 10^4 + 9$ (opposite to hello, almost!)

26. $(177 \times 179 \times 182) + (85 \times 86) - 82$ (two words, good salesman)

27. $(426 \times 474) + (318 \times 487) + 22\,018$ (goes on a hat)

Notes on this activity

This is just a fun activity designed to give the students practice at accurately typing a calculation into their calculators.
Here are the answers.

1. 5317.34 (HE LIES)
2. 505 (SOS)
3. 90439034 (HEDGEHOG)
4. 4509 (GOSH)
5. 378809 (GOBBLE)
6. 938 (BEG)
7. 918, 5508 (BIG BOSS)
8. 7738 (BELL)
9. 7714 (HILL)
10. 317537 (LESLIE)
11. 5318804 (HOBBIES)
12. 379919 (GIGGLE)
13. 35007 (LOOSE)
14. 4519918 (BIGGISH)
15. 77345993 (EGGSHELL)
16. 0.0791 (IGLOO)
17. 55079 (GLOSS)
18. 5537937 (LEGLESS)
19. 53045 (SHOES)
20. 391780 (OBLIGE)
21. 5134, 91805 (HE IS SO BIG)
22. 0.04008 (BOOHOO)
23. 733 (EEL)
24. 35009 (GOOSE)
25. 3180009 (GOODBIE)
26. 5773534 (HE SELLS)
27. 378808 (BOBBLE)

Glossary

Bounds
A value cannot be lower than its lower bound or greater than its upper bound. Bounds are often written as an interval.

Example
If $2 < x \leqslant 5$, x cannot be lower than 2 or greater than 5. However, x cannot equal 2, but it could be 5.

Estimation
An estimation is an approximation of a quantity that has been decided by judgement rather than by carrying out the process needed to produce a more accurate answer. The process might be measuring, doing a calculation or counting.

Examples
An estimation of the number of people in a room might be 30, when the actual measurement (i.e. counting) gives 27.

An estimation of the value of $(23.7 \times 19.1) \div 99.6$ might be 4, or 5, or 4 and a bit. Doing the arithmetic suggests that 4.54 is a good approximation.

Equation
An equation is a statement indicating that two expressions (one of which may be a constant) have the same value.

Examples
$2x + 7 = 15$ and $3(x + 5) = 3x + 15$ are both equations.

Expression
An expression in algebra is a collection of constants and variables that are linked by signs representing operations and which does not include an equals sign.

Examples
$x + y$, $3 + x^2 - y$, $4(x - y)$ and $3x^2 + 5y$ are all expressions.

Rounding
To round a number means to write it to a certain degree of accuracy, rather than writing it exactly. The most common types of rounding are rounding to a certain number of decimal places or a certain number of significant figures.

Significant figures (s.f.)
Significant figures are used to express the relative importance of the digits in a number. The most important digit (the significant figure) is the first non-zero digit, starting from the left-hand side of the number. Starting with the first non-zero digit, all digits are then counted as significant up to the last non-zero digit. After that, zeros may or may not be significant depending on the number of significant figures required.

Examples

	1234.2	**78508**
4 s.f.	1234	78510
3 s.f.	1230	78500
2 s.f.	1200	79000
1 s.f.	1000	80000

	0.024712	**16.317**	**0.0050258**
4 s.f.	0.02471	16.32	0.005026
3 s.f.	0.0247	16.3	0.00503
2 s.f.	0.025	16	0.0050
1 s.f.	0.02	20	0.005

*Note that the number 0.02 has been rounded according to the evidence of the value at the top of its column and not the one immediately above it.

Extension worksheet

This worksheet looks at Goldbach's conjecture and extends it to more than two ways.

General questions

Goldbach's conjecture says that any even number greater than 2 can be made by adding together two prime numbers.

For example:

$4 = 2 + 2$

$6 = 3 + 3$

$8 = 3 + 5$

$10 = 3 + 7$ or $5 + 5$

$12 = 5 + 7$

$14 = 3 + 11$ or $7 + 7$

$16 = 3 + 13$ or $5 + 11$

Answer the following questions.

1. 10 is the first number that can be made in two ways. What is the first number that can be made in three ways? Which is the first number that can be made in four ways? What about five ways?
2. If Goldbach's conjecture is true, why does this mean that any odd number greater than 5 can be made by adding together three primes?

Investigation questions

1. What is a conjecture?
2. Who was Goldbach?
3. Can you find out about any other famous conjectures? What do they say?
4. The number 30 can be made by multiplying the primes 2, 3 and 5.

   Can you make every even number greater than 2 by multiplying primes together? Can any of them be done in more than one way?
5. There are some unsolved maths problems that you can win a million dollars by solving. What are they?

(Answers on page 119)

5 Handling data

This chapter looks at data handling.

Students begin by considering the three averages and the range of a set of data.

There is a substantial exercise on this topic that also covers finding the mean from a frequency table.

Then students look at how data can be presented using:

- pictograms
- stem-and-leaf diagrams
- back-to-back stem-and-leaf diagrams
- pie charts
- tally charts
- two-way tables
- dual bar charts
- composite bar charts
- frequency diagrams (for grouped data).

The difference between discrete and continuous data is then considered before moving on to look at frequency polygons and scatter diagrams, along with the ideas of correlation and a line of best fit.

Prerequisite knowledge

Before attempting this chapter, students should:

- be able to read a scale accurately
- be able to use a calculator
- be able to measure an angle using a protractor
- be able to calculate fractions of quantities.

Investigations, activities and puzzles

An average puzzle

Can you find five numbers with a mean of 4, a median of 2 and a mode of 1?

One possible answer is 1, 1, 2, 5, 11.

Can you find another set of 5 numbers with the same set of averages?

Notes on this activity

This is a good way to get students to think about different kinds of averages. You could also experiment with sets of data that have no mode or are bimodal.

Extension

You can extend this activity by considering the following:

- Are they allowed to use the same number more than once?
- Are they allowed to use zero?
- Are they allowed to use negative numbers?
- Are there any values for the three averages for which you can't find a set of numbers? Why is this?
- Are there any values for the three averages for which the set of numbers is unique?
- What happens if you increase the size of the set of numbers?

What happens to the averages if you:

- multiply all the numbers in the list by 2
- add the same amount to each of the numbers?

How many marbles are in the jar?

Ask students to estimate how many marbles there are in a jar.

When you average out all of their guesses, how accurate is the estimate?

Does your answer become more accurate the more guesses you have?

Notes on this activity

This activity is based on the idea of 'the wisdom of crowds', which says that large groups of people are more intelligent than individuals.

Does this seem to be true in this case?

Average test scores

A student takes two tests. In the first test they score 60% and in the second test they score 90%. Their overall average score is recorded on their report card as 75%, because 75 is the average of 60 and 90.

The student complains and says that they have looked at the two tests and discovered that they actually got 80% of the questions correct.

How has this happened?

Notes on this activity

Many students will not know that it is a bad idea to average percentages.

One way in which the situation described could have arisen is if in the first test they scored 30 out of 50, and in the second test they scored 90 out of 100. The problem is brought about by the fact that the two tests are marked out of two different totals.

Florence Nightingale

Florence Nightingale is mentioned in the introduction to this chapter in the textbook.

Students could do some research and find out about the contributions she made to statistics.

Lengths of words

Find an author who has written books for children and also books for adults.

This will be easy to do as a lot of authors do this.

Look at the first 100 words in one of their adult books and the first 100 words in one of their children's books.

Do they use longer words in the books for adults?

Notes on this activity

This is always an interesting activity.

The results can be made into a composite bar chart.

Alternatively, rather than using two different books written by the same author, you could look at how the same news story has been reported in several different kinds of newspaper.

Glossary

Bar chart
A bar chart is a frequency diagram using rectangles of equal width whose heights or lengths are proportional to the frequency. Usually, adjacent rectangles or bars only touch each other if the data is continuous; for discrete data a space is left between the bars. The bars may be of any width and sometimes are no more than lines.

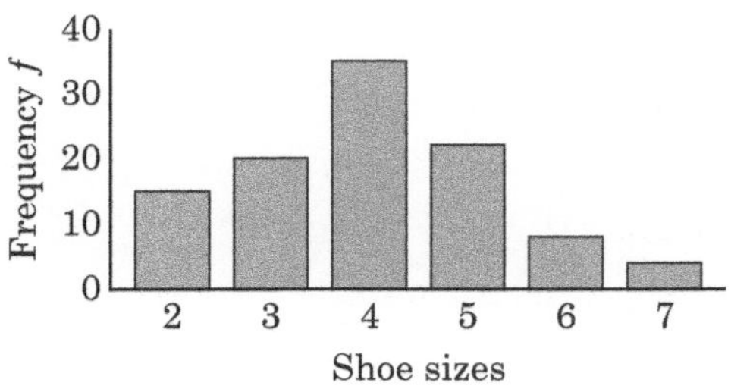

Continuous data
Data is continuous if it can take any value within a given range. Continuous data usually describes things that are measures.

Discrete data
Data is discrete if it can take whole number values. Discrete data usually describes things that are counted.

Grouped data
Grouped data is data that has been organised into intervals that do not overlap to make it easier to process and to spot patterns.

Line of best fit
A line on a graph that shows the general trend of the data points plotted.

Mean
The mean value of a set of data is usually taken to be the arithmetic mean. This is found by adding up the values and dividing the total by the number of values.

Median
The median is the numerical value of the piece of data in the middle of a dataset, after arranging the set in order of size. If there are an even number of pieces of data, the mean of the middle two values is taken as the median.

Example
The dataset 9, 3, 3, 15, 11 has a median of 9 (middle value of 3, 3, 9, 11, 15) whereas 6, 2, 12, 4, 7, 18 has a median of 6.5 (mean of the middle two values, 6 and 7).

Mode
The mode of a set of data is the piece of data that occurs most often.

Example
The dataset 9, 3, 3, 15, 11 has a mode of 3 (as there are more 3s than the other values).

Pictogram
A pictogram is a frequency diagram that uses a symbol to represent a specified number of units of data. The symbol is usually related to the data being shown.

Pie chart
A pie chart is a circular frequency diagram using sectors whose angles at the centre are proportional to the frequency.

Range
The range of a set of numerical data is the numerical difference between the smallest and the greatest values found in that dataset.

Example
For the dataset 9, 3, 3, 15, 11, the range is $15 - 3 = 12$.

Scatter diagram

A scatter diagram shows how two sets of numerical data are related by treating matching pairs of numbers as coordinates and plotting them as a single point. This is repeated for each data-pair.

Examples

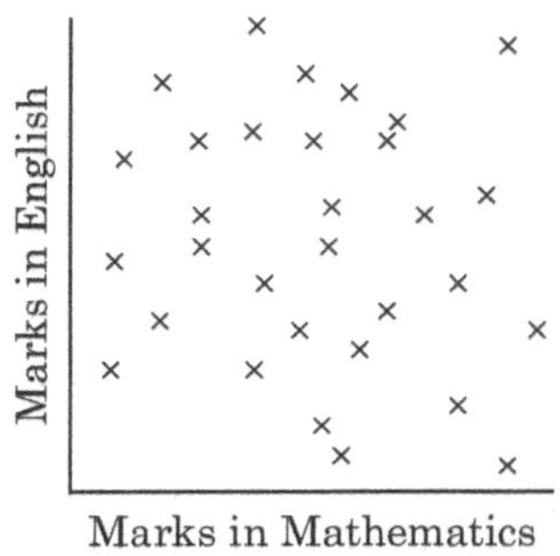

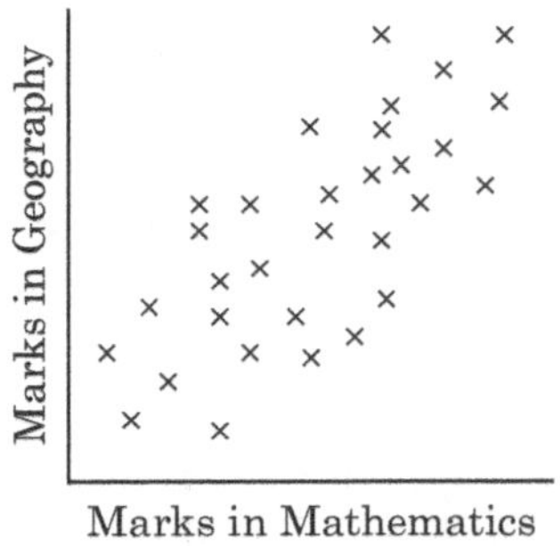

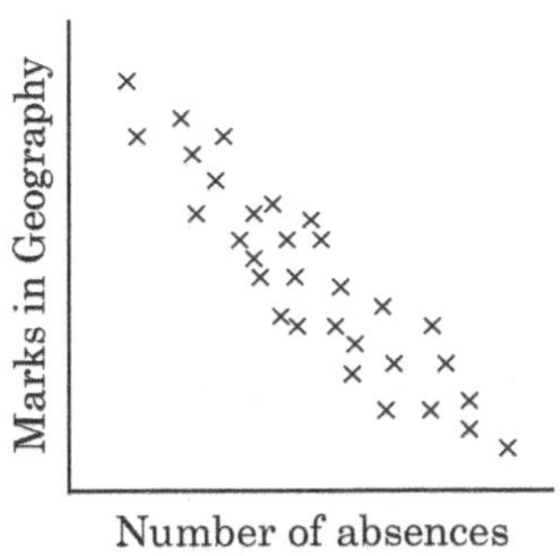

Stem-and-leaf diagram

A stem-and-leaf diagram is a way of organising numerical data into categories according to the first digit (the stem) and subsequent digits (the leaves). Along each row, the leaf numbers are written in ascending order, from left to right. This enables the modal class and the median value to be easily found. A stem-and-leaf diagram must always have a key.

Example

This data set can be organised to form the following stem-and-leaf diagram:

1.5, 2.2, 0.8, 1.0, 2.7, 3.2, 1.8, 2.5, 1.7

0	8			
1	0	5	7	8
2	2	5	7	
3	2			

Key	1 | 0 means 1.0

Tally chart

A tally chart is a way of recording the number of items in each group of a set of data using short vertical lines. Each group of five items of data is represented by four lines with a diagonal line across them.

Two-way table

A two-way table is a table that shows the frequencies of two different categories within a single set of data. The columns show one category and the rows show the other.

Example

A fitness club carried out a survey one morning to find out the ages of the people using the club and the activity they were doing. This two-way table shows the results.

	Gym	Swimming pool	Total
Under 18	5	14	19
18 and over	24	6	30
Total	29	20	49

Extension worksheet

This worksheet looks at finding the mean from a grouped frequency table for continuous data.

It is common to have continuous data displayed in a grouped frequency table like the one below.

Height (cm)	Frequency
$100 < h \leqslant 120$	4
$120 < h \leqslant 140$	6
$140 < h \leqslant 160$	7
$160 < h \leqslant 180$	2
$180 < h \leqslant 200$	1

Since the data has been grouped into classes, the original values have been lost. Nevertheless, an *estimate* of the mean can be calculated by using the *mid-interval values* for each class.

Height (cm)	Frequency	Mid-interval value (MIV)	MIV × frequency
$100 < h \leqslant 120$	4	110	$4 \times 110 = 440$
$120 < h \leqslant 140$	6	130	$6 \times 130 = 780$
$140 < h \leqslant 160$	7	150	$7 \times 150 = 1050$
$160 < h \leqslant 180$	2	170	$2 \times 170 = 340$
$180 < h \leqslant 200$	1	190	$1 \times 190 = 190$
	Total = 20		Total = 2800

The **total** of the last column can then be divided by the **total frequency** (20) to find an estimate of the mean:

$$(440 + 780 + 1050 + 340 + 190) \div 20 = 2800 \div 20 = 140 \text{ cm}$$

Use the above method to find an estimate of the mean for the following sets of grouped data.

1.

Weight (kg)	Frequency
$40 < w \leqslant 50$	2
$50 < w \leqslant 60$	8
$60 < w \leqslant 70$	7
$70 < w \leqslant 80$	3

2.

Time (s)	Frequency
$0 < t \leqslant 20$	5
$20 < t \leqslant 30$	14
$30 < t \leqslant 40$	8
$40 < t \leqslant 60$	3

3.

Height (cm)	Frequency
$80 < h \leqslant 90$	6
$90 < h \leqslant 100$	10
$100 < h \leqslant 120$	18
$120 < h \leqslant 150$	9
$150 < h \leqslant 200$	7

(Answers on page 120)

6 Number 3

This chapter is about the related concepts of fractions, decimals, percentages and ratio.

Skills that are covered are:

- converting fractions into decimals
- finding a percentage of a quantity
- increasing or decreasing a quantity by a percentage
- repeated percentage change
- calculating a percentage change.

Financial applications of percentages that are covered include calculating:

- simple interest
- compound interest
- earnings and tax
- profit.

The chapter then moves on to look at ratio and proportion. This focuses on quantities that are directly proportional, including:

- map scales
- currency exchange
- conversion graphs.

Prerequisite knowledge

Before attempting this chapter, students should be familiar with:

- the idea of a percentage
- fractions and how to calculate with them
- how to divide decimal numbers
- how to read a scale.

Investigations, activities and puzzles

Double discount

Suppose you work in a shoe shop and want to buy a pair of shoes that cost £60.

When you go to buy them you are told that there is currently 15% off this pair of shoes.

However, because you work in the shop, you also get another 20% off as a staff discount.

You can apply the two discounts in either order.

In which order is it best to apply the discounts?

Notes on this activity

This is an interesting puzzle because once the solution is explained, students are often surprised that they didn't see it immediately.

The best way to explain why the order doesn't matter is to describe the percentage decreases in term of multiplying by a decimal.

To reduce the price by 15% you can multiply by 0.85.

To reduce the price by 20% you can multiply by 0.8.

The problem is then choosing between calculating
$60 \times 0.85 \times 0.8$ and $60 \times 0.8 \times 0.85$.

Clearly, they will give the same result, since the order of the numbers in a multiplication does not matter.

Extension

To extend this problem, you could ask what single reduction this double reduction is equivalent to.

They will probably realise that the best way to do this is to combine the two multipliers together to make one single multiplier.

In this case,
$60 \times 0.85 \times 0.8 = 60 \times (0.85 \times 0.8)$
and since $0.85 \times 0.8 = 0.68$,
it is equivalent to a 32% reduction ($1 - 0.68 = 0.32 = 32\%$).

Antipercentages

If a quantity is increased by 25%, by what percentage must the new quantity be decreased in order to get it back to its original size?

Notes on this activity

The answer in this case is 20%, so we can say that 20% is the antipercentage of 25%.

Students will often immediately assume that the answer will be 25% because they do not realise that a percentage is always a percentage of some amount and therefore will change in size depending on how big that amount is.

See whether students can work out some other pairs of percentages and antipercentages.

For example, if a quantity is increased by p per cent, to return it to its original size you would need to reduce the new quantity by $\frac{100p}{100+p}$ per cent.

This can be shown in the following way.

If the original quantity is q and we increase this by p per cent, it will then become

$$q \;\; 1+\frac{p}{100}$$

So, what must we multiply the new amount by to get back to q?

The answer is the reciprocal of $1+\frac{p}{100}$ which is $\frac{100}{100+p}$

This is equivalent to $\frac{100+p-p}{100+p}=\frac{100+p}{100+p}-\frac{p}{100+p}=1-\frac{p}{100+p}$

Multiplying by this number will decrease a value by $\frac{p}{100+p}=\frac{100p}{100+p}$ per cent.

Is there an easier way to show this algebraically?

Extension

So far, we have only been considering how you reverse a percentage increase.

How would you reverse a percentage decrease?

To reverse a percentage decrease, will you need to increase the new amount by a percentage that is bigger or smaller than the percentage you decreased the original amount by?

Glossary

Compound interest

When an amount of money is invested in an account that pays compound interest, the compound interest earned at the end of each period is worked out on the principal (the amount invested initially) plus any previous interest already earned. The calculation gives the total amount in the account, rather than just interest earned.

total amount in account =

$$\text{principal} \times \ 1 + \frac{\text{rate of interest (\%)}}{100}^{\ \text{number of periods}}$$

Example

If \$300 is invested in an account that pays compound interest at the rate of 2% per year, then the total amount of money in the account at the end of 4 years would be

$$\$300 \times \ 1 + \frac{2}{100}^{\ 4} = \$324.73$$

Conversion graph

A conversion graph is a graph used to convert between different units. For example, it could be a graph used to convert between degrees Fahrenheit and degrees Celsius.

Direct proportion

When two sets of values are related using a constant multiplier (so that an increase in one matches an increase in the other) they are said to be in direct proportion (or directly proportional). This is also known as direct variation.

$\propto$ is the symbol for 'is proportional to' and k is used for the constant.

$y \propto x$ means that y is directly proportional to x; or $y = kx$.

Example

The perimeter of a square is directly proportional to the side length of the square because it is always exactly four times the length of one side, however big the square is.

This can be displayed on a graph.

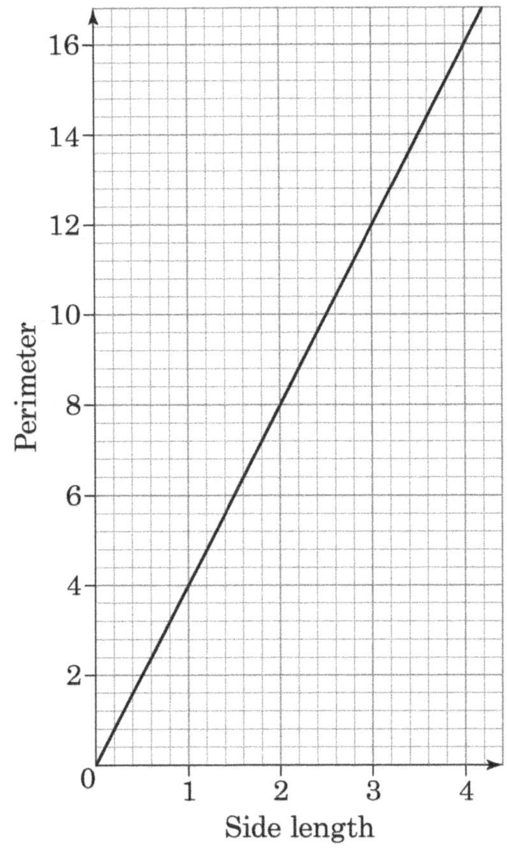

Note that graphs of quantities that are directly proportional will always be straight lines that pass through the origin.

Inverse proportion

When two sets of values are related and an increase in one matches a decrease in the other, they are said to be in inverse proportion (or inversely proportional). This is also known as indirect variation.

$y \propto \frac{1}{x}$ means that y is inversely proportion to x; or $y = \frac{k}{x}$, where k is a numerical constant.

Example

The sets {40, 24, 15} and {3, 5, 8} are in inverse proportion to each other since

$40 \times 3 = 120$, $24 \times 5 = 120$ and $15 \times 8 = 120$.

Per cent

A value given as a 'per cent' means that the number stated can be used to make a fraction with that number on the top and 100 on the bottom. % is the symbol for per cent.

Examples

45 per cent is the fraction $\frac{45}{100}$;

37% is 37 per cent or $\frac{37}{100}$

Ratio

This is used to compare the sizes of two (or more) quantities.

Example

Mortar for building a brick wall is made by mixing 2 parts of cement to 7 parts of sand. (The parts may be decided by weight or by volume, as long as the same units are used.) Then it can be said that the ratio of cement to sand is 2 to 7 which is also written in the form 2 : 7.

Simple interest

Simple interest is interest worked out at the end of each period and is always calculated on only the principal (the original amount).

Example

A loan of \$300 at an interest rate of 2% per month simple interest would mean that, at the end of each month, interest of $0.02 \times \$300 = \$6$ would be owing. So, at the end of a year, the total interest owing would be $12 \times \$6 = \$72$ (plus the loan of \$300 which still has to be repaid).

total simple interest =

$$\frac{\text{principal} \times \text{rate of interest (\%)} \times \text{number of periods}}{100}$$

Extension worksheet

This worksheet extends the work on compound interest to consider how being paid interest more frequently affects the amount of interest earned.

If you invest $1 in a savings account that pays compound interest at a rate of 100%, how much would you have after 1 year?

The answer is $2.

Banks, however, do not always calculate interest once a year. They may spread the interest out and calculate it more regularly.

For example, rather than paying 100% interest once a year, they may pay 50% twice a year.

If they do this, you will earn more interest because increasing by 50% twice gives you more than increasing by 100% once.

$$\$1 \times 1.50 = \$1.50$$

$$\$1.50 \times 1.50 = \$2.25$$

So you will earn $2.25 instead of $2, which is $0.25 more than if you were paid 50% compound interest twice.

Would you earn even more if you were paid 25% four times a year?

$$\$1 \times 1.25 = \$1.25$$

$$\$1.25 \times 1.25 = \$1.56$$

$$\$1.56 \times 1.25 = \$1.95$$

$$\$1.95 \times 1.25 = \$2.44$$

Yes! You would earn $2.44 if you were paid 25% four times a year, which is $0.19 more than if you were paid 50% twice a year.

What if you were paid 20% five times a year or 10% ten times a year? Would the amount you are paid keep going up?

If the interest were calculated every hour, would you earn more than if it were calculated every day?

Is there a limit to how much you can earn in one year?

(Answers on page 121)

7 Shape and space 2

This chapter is mostly about lengths and areas and their associated units.

It starts by setting the scene with a study of the metric units related to length, area, volume and mass. Conversion rules are considered, along with how these rules change when the units are two or three dimensional.

Units of time are then considered and intervals of time are measured.

After that, the focus turns to compound units.

The main compound unit considered is speed, but others are also considered, such as pressure, density and population density.

Metric units are then used to calculate areas and perimeters of:

- rectangles
- triangles
- compound shapes
- parallelograms
- trapezia.

The circumference and area of a circle are then studied, along with the related concepts of arcs and sectors.

The chapter closes by looking at volume and surface area of:

- cuboids
- prisms
- cylinders
- cones
- pyramids
- spheres.

Prerequisite knowledge

This chapter assumes a certain degree of familiarity with the basic units being used, along with the ability to use a calculator.

Investigations, activities and puzzles

A trip to the seaside

A family take a trip to the seaside.

On the way, their average speed is 30 km/h.

On the way back, they encounter traffic and their average speed is only 20 km/h.

What is their average speed for the whole journey?

Notes on this activity

This is a very well-known puzzle. Most people's first instinct is to suspect that the answer is 25 km/h, but the answer is actually 24 km/h. This is because they spend more time travelling at the slower speed than at the faster speed and what is important is how long they are travelling at that speed rather than how many miles they travel at that speed.

An interesting feature of the puzzle is that the actual distance to the seaside is not important. When you do the algebra you will see that the distance cancels out. This means that if the students are struggling to solve the problem you can tell them to make up a distance, so that they can use it in their calculations. A sensible distance to use is 60 km.

If you suppose that the distance is 60 km, you get the following:

time = distance ÷ speed

So,

Time to get to the seaside = 60 ÷ 30 = 2 hours.

Time to get back home = 60 ÷ 20 = 3 hours.

Total distance = 2 × 60 = 120 km

Total time spent travelling = 5 hours

Average speed for the whole journey = 120 ÷ 5 = 24 km/h

However, if you don't substitute a value for the distance to the seaside and instead call it d, you get the following:

Time to get to the seaside = $\frac{d}{30}$ hours.

Time to get back home = $\frac{d}{20}$ hours.

Total distance = $2d$ km

Average speed for the whole journey =

$$\frac{\text{total distance}}{\text{total time}} = \frac{2d}{\frac{d}{30} + \frac{d}{20}} = \frac{2d}{\frac{2d + 3d}{60}} = \frac{2d}{\frac{5d}{60}} = \frac{2\cancel{d}^{1} \times 60}{5\cancel{d}^{1}} = \frac{120}{5} = 24 \text{ km/h}$$

Area and perimeter

1. Consider these two rectangles.

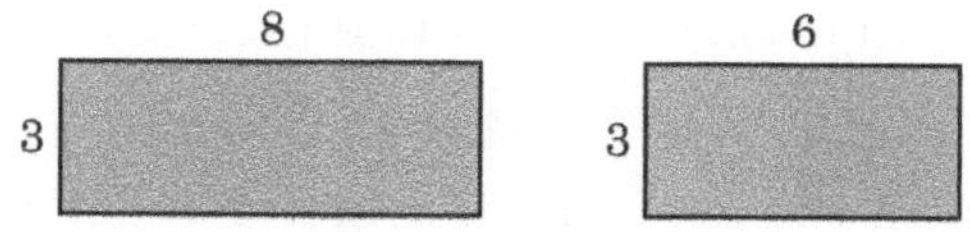

The first rectangle has a perimeter of 22 units and an area of 24 square units.

The second rectangle, however, has a perimeter of 18 units and an area of 18 square units.

How many rectangles can you find whose perimeter and area have the same numerical value?

2. Extend this idea to consider rectangles with pieces removed.

How does this affect your results?

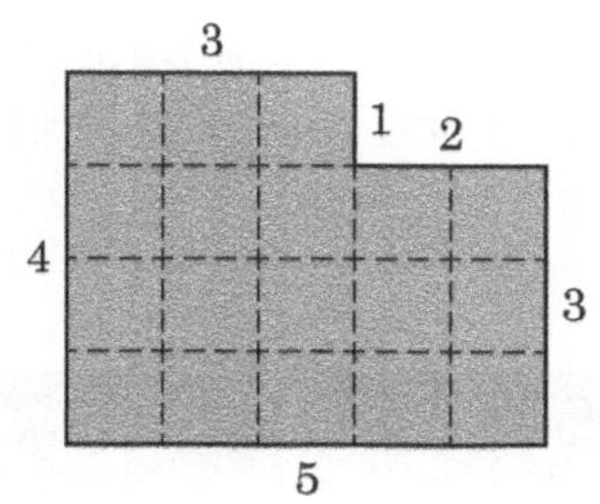

Notes on Question 1 of this activity

When you analyse the first part of this activity you find the following.

If a rectangle has a length a and a width b, the area is ab and the perimeter is $2(a + b)$.

If the perimeter and area have the same numerical value then:

$$2(a + b) = ab$$
$$2a + 2b = ab$$
$$ab - 2a = 2b$$
$$a(b - 2) = 2b$$
$$a = \frac{2b}{b-2}$$

which means that a will be a whole number when $b - 2$ is a factor of $2b$. This should help students find solutions to the problem.

Notes on Question 2 of this activity

Note that when you remove squares from the corner of a rectangle you don't actually change the perimeter, even though you do change the area.

This means that if the area and perimeter are not equal, as long as the original area of the rectangle is greater than the perimeter, you may be able to just remove parts of the area until you reach the required total. For example, in the diagram that accompanies Question 2, notice that the original 4 by 5 rectangle would have had an area of 20, and its perimeter would be only 18. You can therefore remove two squares from the top corner to reduce the area by 2, while not changing the value of the perimeter.

The activity for the students then becomes trying to work out to what extent this approach will work. What are its limitations? Are we going to allow the rectangle to be decreased by certain amounts? For example, are you only allowed to remove another rectangle? Or can you remove any number of squares? At what point does this activity cease to be interesting? This in itself might be an interesting question to consider.

Extension

You could extend this activity even further by considering the following:

- What happens if you do this with right-angled triangles or equilateral triangles?
- Could this idea work with a semi-circle?
- How about three-dimensional shapes? Can you make cuboids, spheres, or cylinders whose volumes are numerically equal to their surface area?
- Can you find any connection between the square with equal area and perimeter and the circle with equal area and perimeter? How about the equilateral triangle with equal area and perimeter?

Square units

Can any of your students write their name in a square whose size is $0.01\,m^2$?

Notes on this activity

When you ask students this question, you will probably find that most of them attempt to write their name inside a square of side length 1 cm. This is because they do not realise how the unit conversion rules change when you are dealing with square units.

($1\,m^2 = 100 \times 100 = 10\,000\,cm^2$, so $0.01\,m^2 = 100\,cm^2$)

Glossary

Arc

An arc of a circle is any piece of the curve that makes the circle.

$$\text{length of arc} = \frac{\text{angle (in degrees) of arc at centre}}{360} \times \text{circumference of full circle}$$

Average speed

When an object moves over some distance its speed may vary as it travels. The average speed is found by considering the total time taken for the movement and the total distance moved. The three equations used when working with average speed are:

speed = distance ÷ time
time = distance ÷ speed
distance = speed × time

Example

A car is driven from Aswan to Cairo, a distance of 780 km, and the journey takes 20 hours. What is the average speed for the journey?

The average speed is 780 ÷ 20 = 39 km/h.

Circle

A circle is either a closed curve (i.e. a line that curves around and joins up with itself) in a plane (a flat surface), which is everywhere the same distance from a single fixed point, or it is the shape enclosed by that curve.

area of circle = π × radius × radius = πr^2, where the radius is half of the diameter,

i.e. $r = \frac{d}{2}$

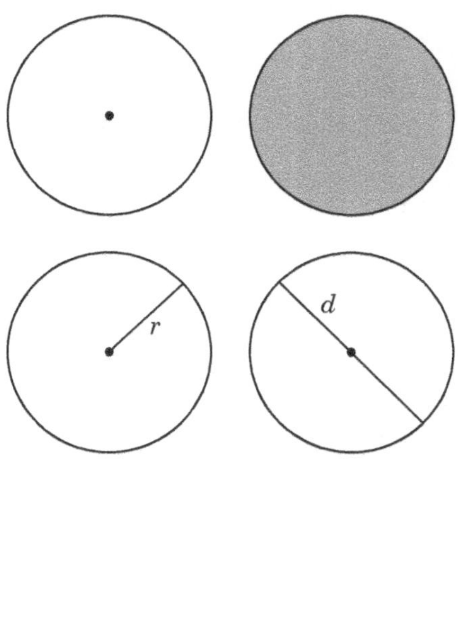

Compound shape

A shape made by joining two or more other shapes together.

For example, this is a compound shape made from a rectangle and a semi-circle.

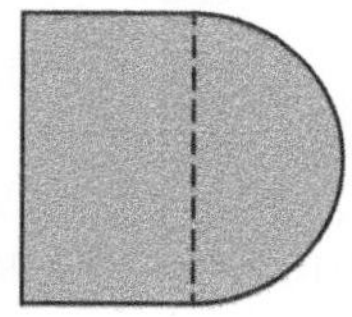

Cone

A cone is the three-dimensional shape formed by a straight line when one end is moved around a circle, while the other end of the line is kept fixed at a point which is not in the plane of the curve.

volume = $\frac{1}{3}\pi r^2 h$

surface area
= area of circular base + area of curved surface
= $\pi r^2 + \pi r l$

where l is the slant height of the cone.

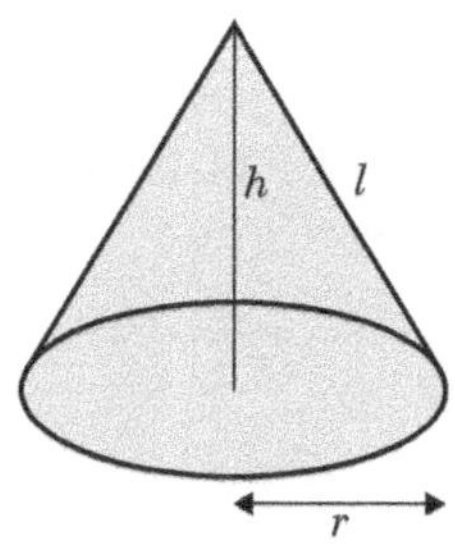

Cuboid

A prism whose cross-section is a rectangle.

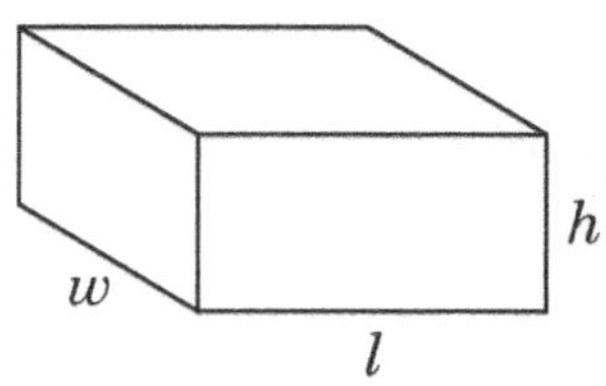

volume = length × width × height
= $l \times w \times h$

surface area = $2 \times (wl + hl + wh)$

Cylinder

A prism with a circular cross-section.

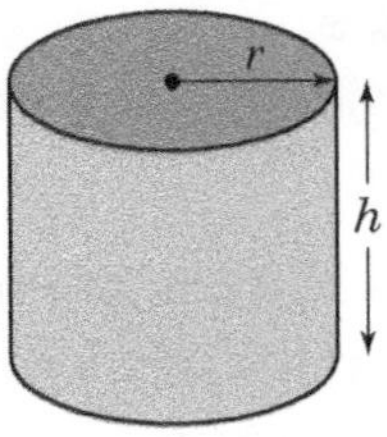

volume = $\pi r^2 h$

Prism

An object with a uniform cross-section.

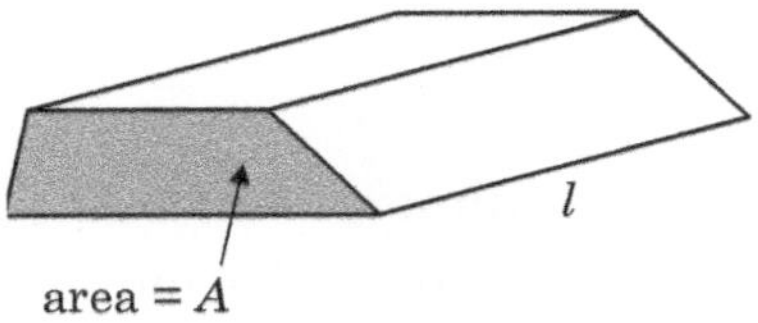

volume = $A \times l$

Pyramid

In a pyramid, every point on the perimeter of the base is connected to a single point at the top.

The base of a pyramid is usually a square but it does not have to be.

The volume of a pyramid = $\frac{1}{3} \times$ area of base $\times$ perpendicular height.

There is no general formula given for the surface area of a pyramid.

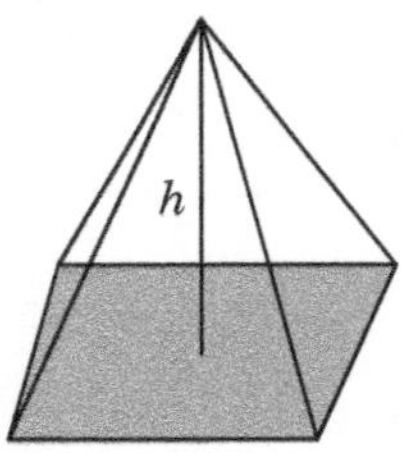

Rectangle

A quadrilateral whose angles are all right angles.

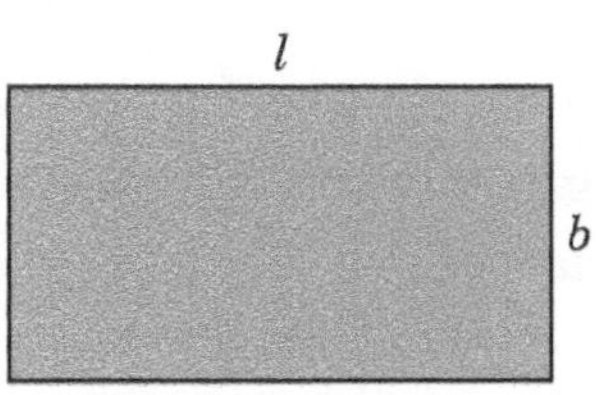

area = $l \times b$

Sector

A sector of a circle is the shape enclosed between an arc and the two radii at either end of that arc.

area of sector =

$$\frac{\text{angle (in degrees) of arc at centre}}{360} \times \text{area of full circle}$$

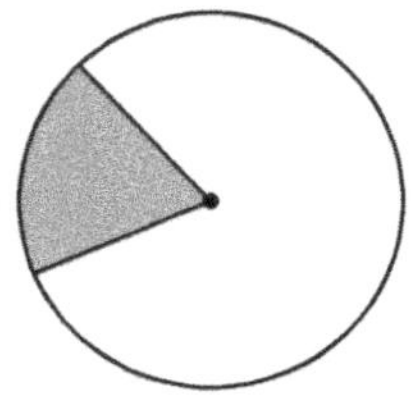

Sphere

A sphere is either the shape of a surface in three dimensions that is everywhere the same distance from a single fixed point or it is the solid shape enclosed by that surface. The balls used to play most games are spheres.

volume = $\frac{4}{3} \neq r^3$

surface area = $4\pi r^2$

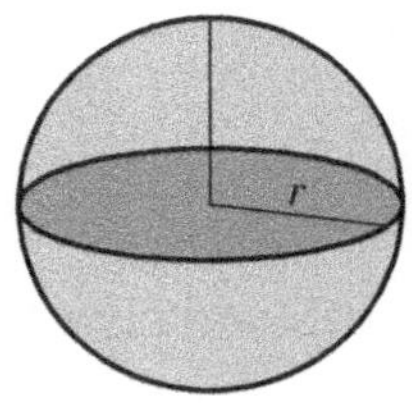

Surface area

The surface area of a solid is the total area taken up by its surface.

Trapezium

A quadrilateral with two parallel sides.

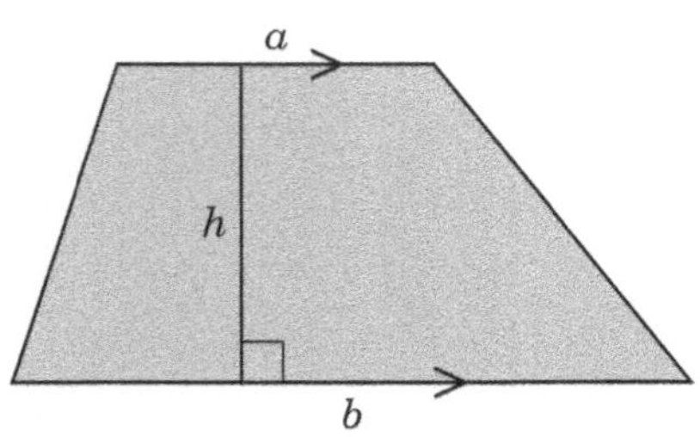

area = $\frac{1}{2}(a + b) \times h$

Volume

The volume of a solid is the amount of three-dimensional space it takes up.

Extension worksheet

This worksheet extends the work on calculating volume to include the use of similar triangles.

If you take a cone and slice the top off, parallel to the base, you create another mathematical shape called a **frustum**.

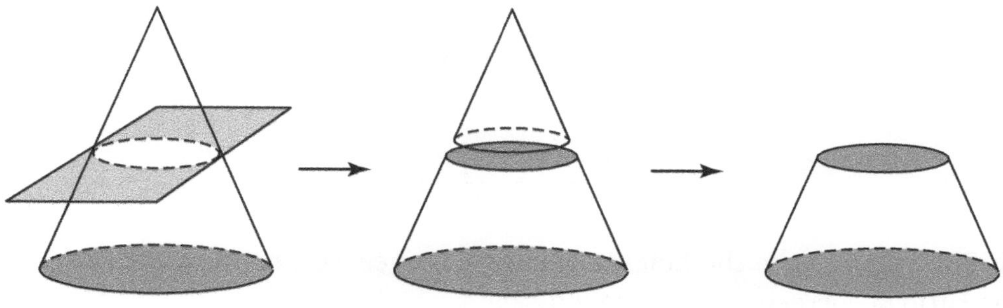

R_1 is the radius of the base of the original cone.

R_2 is the radius of the small cone you have sliced off.

h is the height of the frustum, that is, the piece that remains.

To find the volume of the frustum, you have to find the volume of the original cone and subtract from this the volume of the small cone that has been removed.

Example

This frustum is formed by taking a cone of base radius 6 cm and slicing off a cone of base radius 3 cm.

To find the volume of a cone you need to use the formula $V = \frac{1}{3} \neq r^2 h$, where h is the height of the cone.

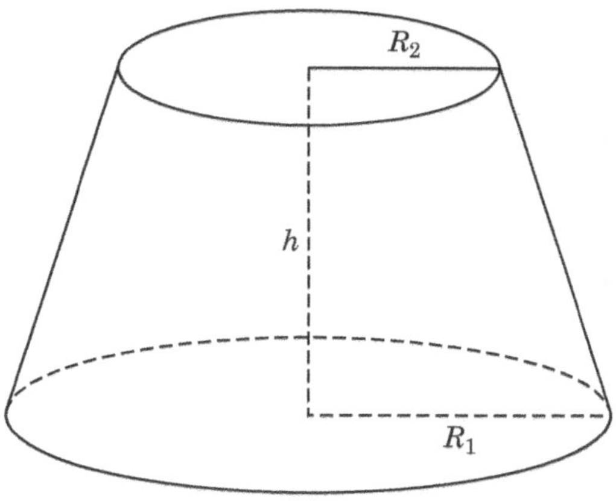

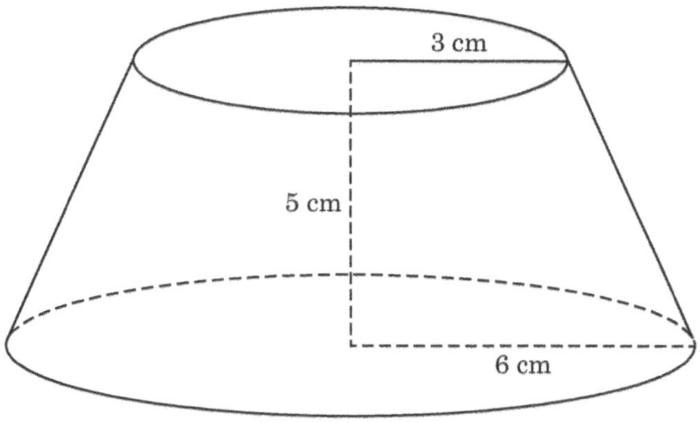

To find the height of the original cone, and the cone that has been removed, you need to use similar triangles.

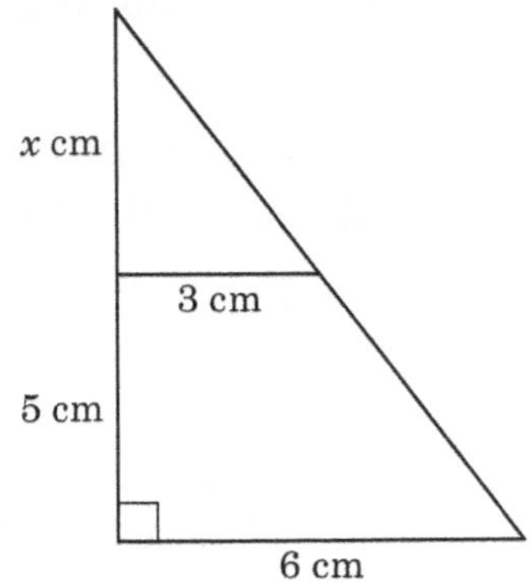

To find x and hence the heights of the two cones, set up an equation for the ratio of corresponding side lengths:

$$\frac{6}{3} = \frac{5+x}{x} \qquad 6x = 3(5+x)$$

Now solve this equation to find x: $6x = 15 + 3x \quad 3x = 15 \quad x = 5$

So, the height of the original cone is 10 cm and the height of the small cone that has been removed is 5 cm.

The volume of the frustum is therefore:

$$V = \frac{1}{3}\pi \times 6^2 \times 10 - \frac{1}{3}\pi \times 3^2 \times 5 = 105\pi = 330\text{cm}^3 \text{ (3 s.f.)}$$

Find the volume of these frustums, giving your answers correct to 3 significant figures.

1. $R_1 = 4$ cm, $R_2 = 1$ cm, $h = 3$ cm
2. $R_1 = 8$ cm, $R_2 = 4$ cm, $h = 6$ cm
3. $R_1 = 10$ cm, $R_2 = 2.5$ cm, $h = 2$ cm
4. $R_1 = 20$ cm, $R_2 = 4$ cm, $h = 7$ cm
5. $R_1 = 11.5$ cm, $R_2 = 5.75$ cm, $h = 4.5$ cm

(Answers on page 122)

8 Algebra 2

This chapter is about graphs.

In the course of the chapter, students will learn about:

- straight line graphs, focusing on
  - gradients and intercepts
  - equations of the form $y = mx + c$
- quadratic graphs
- reciprocal graphs.

Students then look at how graphs can be used to solve equations.

They also look at how graphs can be used to describe real-life situations, including travel graphs.

Prerequisite knowledge

This chapter assumes that the students are familiar with two-dimensional Cartesian coordinates.

Investigations, activities and puzzles

Defining a line

If you know the gradient of a straight line and also its y-intercept, then you know exactly which line is being referred to. Are there any other ways to uniquely define a straight line?

Notes on this activity

Here are some examples of what students might say:

- two points on the line
- the two intercepts
- the gradient of the line and any point on the line.

You could ask students what the quickest way to find the equation of a line would be if you were given each of these pieces of information.

Extension

Some possible extensions are:

- Does this apply to quadratic curves? How many pieces of information do you need to know in order to define a quadratic curve?
- What about other kinds of curves?

This opens up an interesting investigation regarding how many pieces of information are needed to define various different mathematical objects.

Some objects require just one piece of information. For example:

- a circle (its radius)
- a square (its side length).

A triangle requires three pieces of information to uniquely define it:

- angle, side, angle (ASA) or
- side, angle, side (SAS) or
- side, side, side (SSS) or
- right angle, hypotenuse, side (RHS).

Which shapes require two pieces of information to be accurately defined?

One example is a rectangle. Can the students think of any others?

To uniquely define a circle *and* its position you need one extra piece of information: the coordinates of its centre point. How many extra pieces of information are needed to uniquely define a square and its position? What about the other shapes mentioned above?

Notes on this activity

Three points are required to uniquely define a quadratic curve. This is because a general quadratic curve is of the form $f(x) = ax^2 + bx + c$. To determine the exact curve means finding three values a, b and c. This therefore requires three equations.

We can deduce from this that if you raise the order of the equation to, say, a cubic of the form $f(x) = ax^3 + bx^2 + cx + d$, then four points would be required to determine the four coefficients.

Enclosed spaces

The three lines $y = 1$

$y = x$

$y = 6 - x$

enclose an area of 4 square units.

Can you find the equations of three straight lines that enclose an area of 10 square units?

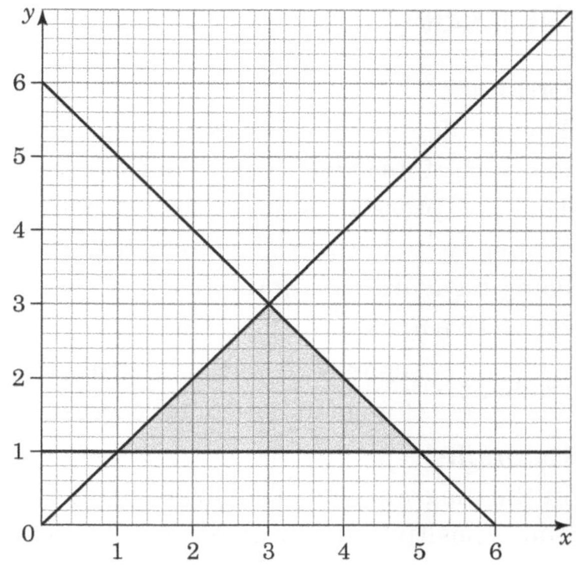

Notes on this activity

This involves students working out the equations of several different straight lines as well as using the formula for the area of a triangle.

Students could work systematically through the numbers from 1 to 10 and try to find sets of lines that enclose those areas. Can they spot any patterns?

Notes on this activity

It is easier to work out the enclosed area by subtracting triangles from a rectangle, rather than trying to directly work out the enclosed area.

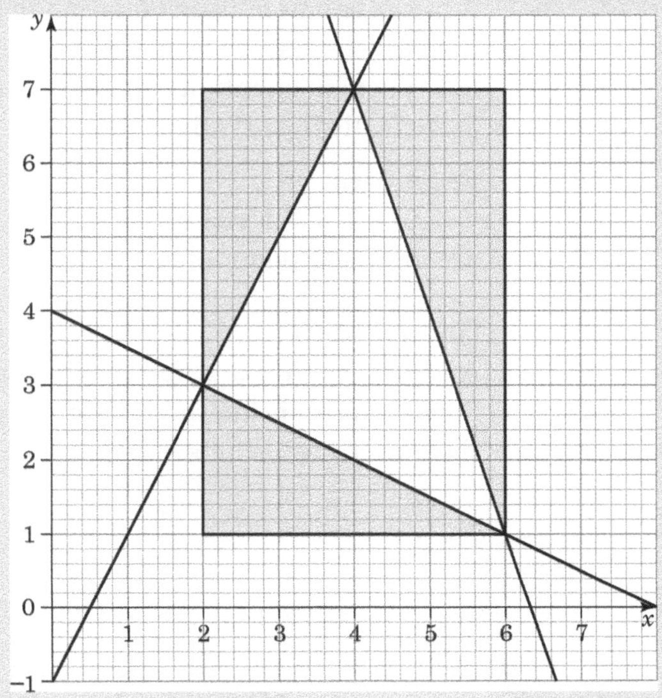

Extension

Can students work out how to enclose a given area using three lines, none of which are horizontal or vertical?

For example, the lines $y = -\frac{1}{2}x + 4$, $y = 2x - 1$ and $y = 19 - 3x$ enclose an area of 10 square units.

Glossary

Gradient

The gradient of a line drawn on a graph is a measure of its slope relative to the x-axis. This is expressed as the ratio of its vertical change to its horizontal change, both changes being measured using the scales of their respective axes.

$$\text{gradient} = \frac{\text{change in vertical coordinate}}{\text{change in horizontal coordinate}}$$

Example

In the diagram, the gradient is given by $a \div b$ and is positive since y increases as x increases.

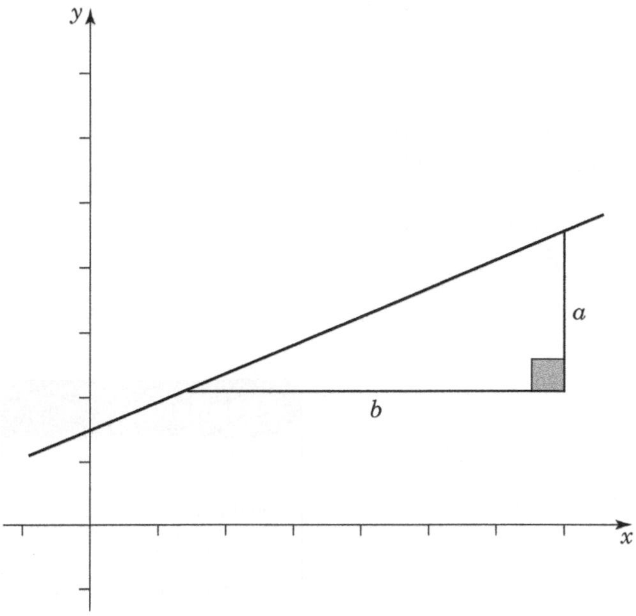

In the diagram, the gradient is given by $a \div b$ and is negative since y decreases and x increases.

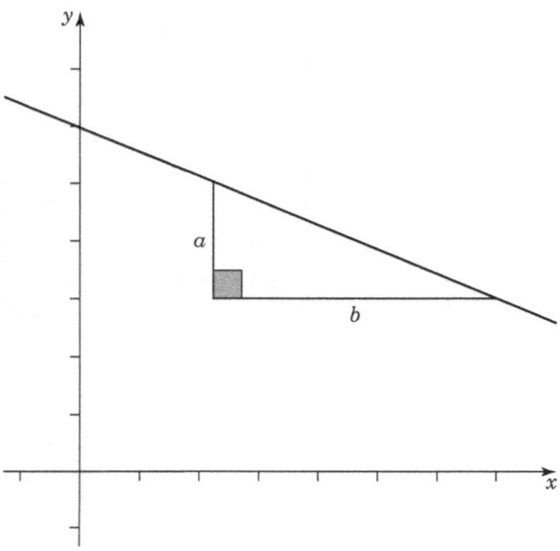

Intercept

The intercept of a graph is the point at which it cuts across an axis. For linear graphs, this word usually refers to the point at which the line cuts the y-axis (y-intercept).

Example

In the linear graph shown, the y-intercept is -2.

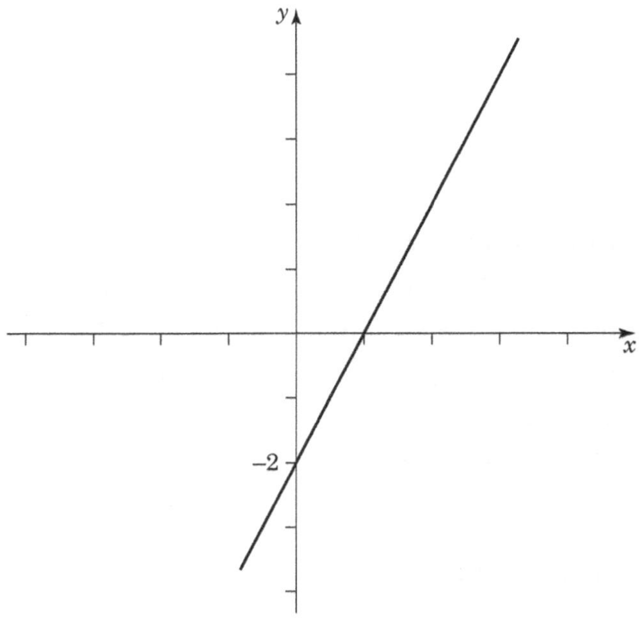

Linear equation

A linear equation is an equation in which the highest power (or index) of any variable is 1. Such an equation can be represented graphically by a straight line.

Examples

$y = 3x + 2$, $y = 4$ and $x = 3y - 5$ are all linear equations.

Linear graph

A linear graph is a graph in which all the points representing the relationship between the quantities lie on a straight line.

Quadratic graph

A quadratic graph is a graph in which the relationship between the variables is given by a quadratic equation. Its shape is called a parabola.

Examples

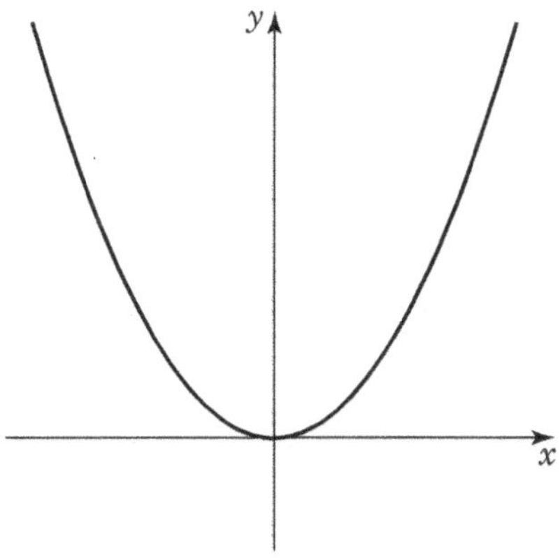

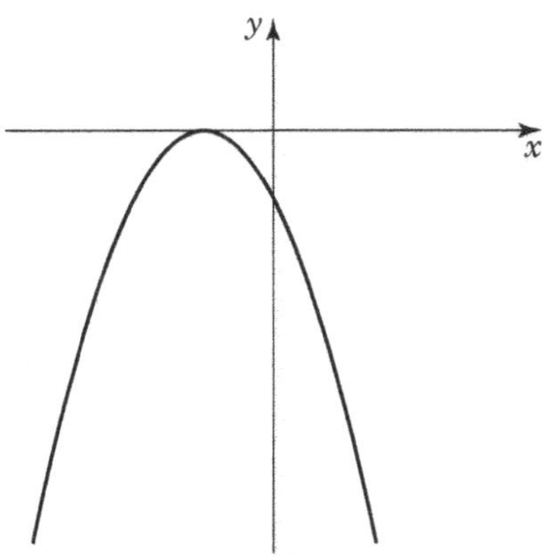

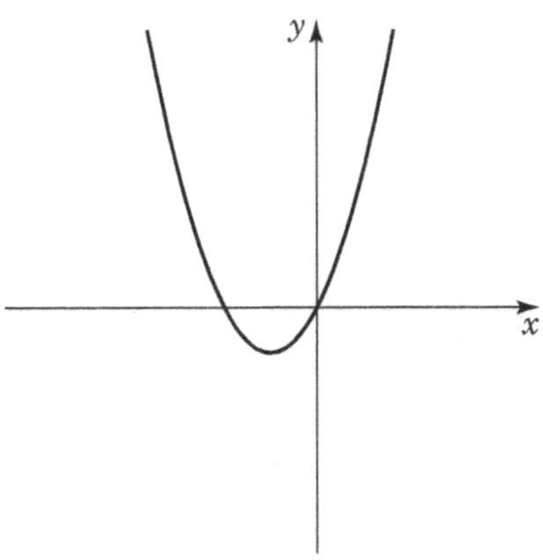

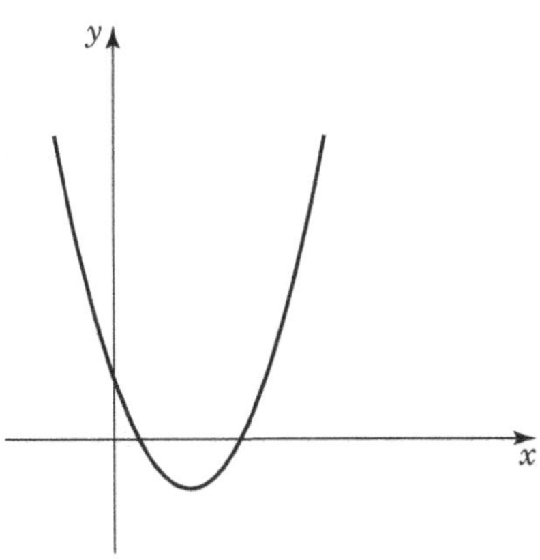

Travel graphs

These can be distance–time graphs or velocity–time graphs. In both of these graphs, time is always on the horizontal axis. The gradient of a distance–time graph represents the velocity. The gradient of a velocity–time graph represents the acceleration.

Example

This graph shows a journey by a cyclist.

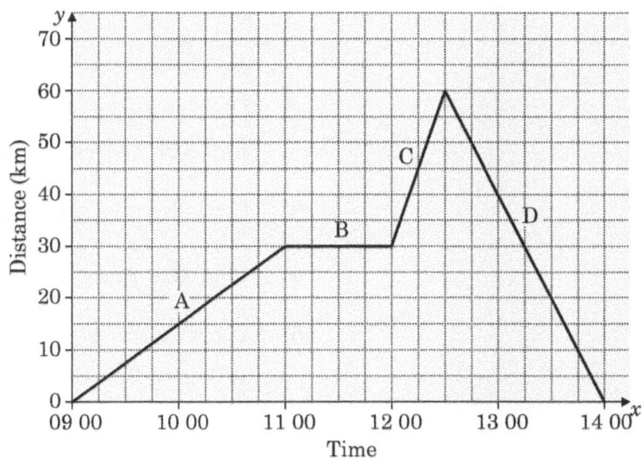

Extension worksheet

This worksheet develops the method of graphical solutions to equations by including examples of cubic equations and reciprocal graphs.

You have seen graphs of quadratic functions and have used these to solve quadratic equations. Other types of equations can be solved in the same way.

1. Complete the table of values for the cubic function $y = x^3 - 2$.

x	–3	–2	–1	0	1	2	3
y							

2. Plot the graph using a scale of 1 cm per unit on the x-axis and 2 cm per five units on the y-axis (plot y from 25 to –30).

3. Use the graph to solve the equation $x^3 - 2 = 0$.

4. Use the graph to solve the equation $x^3 - 2 = 6$.

5. What line would have to be drawn on the graph to solve the equation $x^3 - 2 = x$?

   Draw the line and find the approximate solution to the equation.

6. Repeat the method above to solve the following equations:

   a) $x^3 - x^2 + 3 = 0$

   b) $x^3 - x^2 + 3 = 5$

   c) $x^3 - x^2 + 3 = x$

7. Complete the following table of values for the function $y = \frac{1}{x}$.

   Explain why there is no value given for y when $x = 0$.

x	0	0.1	0.4	0.5	1	2	3
y							

8. Plot the graph using a scale of 1 cm per unit on both the x-axis and the y-axis (plot y from 0 to 10).

9. Use the graph to solve the equation $\frac{1}{x} = 3$.

10. Use the graph to solve the equation $\frac{1}{x} = 1.5$.

11. By adding an appropriate extra line to the graph, solve the equation $\frac{1}{x} = x + 1$.

(Answers on page 124)

9 Probability

In this chapter, probability is taught from the very beginning, as though the students have never encountered it before.

It starts with a discussion of what probability is, describing in words how likely an event is to happen, before moving on to expressing a probability as a number between 0 and 1.

The P notation is introduced to mean 'the probability of', and simple examples using balls, dice, discs, cards, spinners and sweets are presented.

The more complicated concepts of mutually exclusive events and relative frequency are then introduced.

The second half of the chapter looks at diagrams that can be used to help to solve probability problems. These are:

- Venn diagrams
- sample space diagrams
- tree diagrams.

Prerequisite knowledge

This chapter assumes that the students are familiar with:

- two-way tables
- fractions
- basic items commonly used in probability questions, such as
  - coins
  - dice
  - playing cards
  - spinners.

Investigations, activities and puzzles

Estimating pi (π)

You can use relative frequency to estimate the value of π.

On a fairly large piece of paper, draw a quarter circle inside a square.

Drop drawing pins (or something else that's quite small) randomly onto the paper.

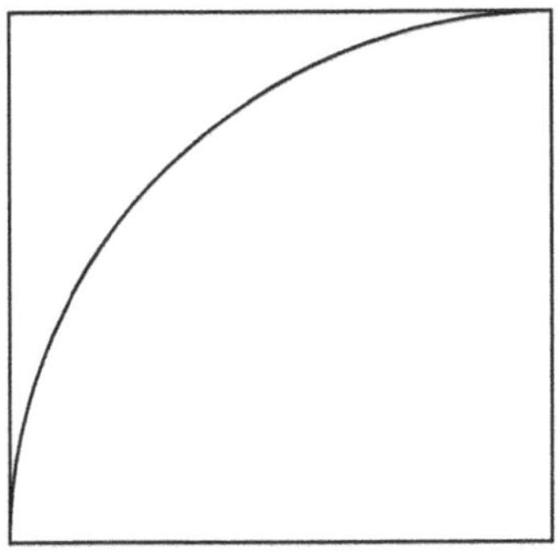

Work out an estimate of the probability that a randomly dropped pin will land in the quarter circle using relative frequency.

You know that if the quarter circle has a radius of r, then the area of the square is r^2.

The area of the quarter circle is $\frac{1}{4}\pi r^2$.

This means the probability of a pin landing in the quarter circle should be $\frac{1}{4}\pi$.

Multiply your relative frequency by 4 to obtain an estimate of π.

How accurate was your estimate?

Notes on this activity

This is called the Monte Carlo method for estimating π and it is well documented online.

It is important that students realise that in order for the method to work, the pins really do need to be randomly dropped onto the paper. If this method fails badly, it is probably due to a lack of randomness in the pin dropping, rather than in the method itself.

The birthday problem

How big does a group of people need to be for the probability of two of them sharing the same birthday to be less than a half?

Notes on this activity

It is well known now among teachers of mathematics that the answer to this question is that you only need 23 people in a group for there to be more than a 50% probability of two of them having the same birthday.

Familiarity with this result, though, does not make it any less surprising or counterintuitive.

One good thing about this is that it can be tested quite easily at school, since most classes contain more than 23 students. It is always good to start by asking students how many people they think you would need in a group to guarantee that two of them share a birthday. The answer is, of course, 366 (if you base it on there being 365 possible birthdays). This question acts as misdirection, as well as being a good illustration of the pigeonhole principle. When you then ask how many people you would need to bring the probability down from certain to 50%, students may expect the answer to be about half of 365. The answer of 23 will always come as a surprise to anyone who has not heard this before.

The way to see that only 23 people are required in order for the probability of two of them sharing a birthday to be more than 0.5 is to think about the problem backwards.

For any number of people, n, what is the probability that their birthdays are all different?

Given a standard 365 day year, we obtain the following result:

The probability of all of them having a different birthday is

$$\frac{365}{365} \times \frac{364}{365} \times \frac{363}{365} \times \cdots \times \frac{365-(n-1)}{365}$$

The probability of at least two of them sharing a birthday is 1 minus this value.

We can set up a spreadsheet to track what happens to this as n increases.

n	P (all different)	1 − P (all different)
1	1	0
2	0.997260274	0.002739726
3	0.991795834	0.008204166
4	0.983644088	0.016355912
5	0.972864426	0.027135574
6	0.959537516	0.040462484
7	0.943764297	0.056235703
8	0.925664708	0.074335292
9	0.905376166	0.094623834
10	0.883051822	0.116948178
11	0.858858622	0.141141378
12	0.832975211	0.167024789
13	0.805589725	0.194410275
14	0.776897488	0.223102512
15	0.74709868	0.25290132
16	0.716395995	0.283604005
17	0.684992335	0.315007665
18	0.653088582	0.346911418
19	0.620881474	0.379118526
20	0.588561616	0.411438384
21	0.556311665	0.443688335
22	0.524304692	0.475695308
23	0.492702766	0.507297234
24	0.461655742	0.538344258
25	0.431300296	0.568699704
26	0.40175918	0.59824082
27	0.373140718	0.626859282
28	0.345538528	0.654461472
29	0.319031463	0.680968537
30	0.293683757	0.706316243

From this, you can see that 23 people is the point at which the probability becomes greater than 0.5 for the first time.

Extension

You can extend this problem by changing the percentage and asking students to work out how many people you would need for the probability to be your chosen value.

Extension

Bertrand's box paradox

You have three boxes:

- Box 1 contains two gold coins.
- Box 2 contains two silver coins.
- Box 3 contains one gold coin and one silver coin.

If you choose a coin at random from one of the boxes and it is a gold coin, what is the probability that the other coin in the box is also gold?

Notes on this activity

Bertrand's box paradox was first posed by the French mathematician Joseph Bertrand in 1889. It is quite commonly used when teaching probability because it is a simple example of a counterintuitive result that will make students think more deeply when they first study probability.

The answer to the question is $\frac{2}{3}$, although at first some students may jump to the conclusion that it must be $\frac{1}{2}$, since if a gold coin is chosen, there are only two options: either the other coin is gold or the other coin is silver.

The reason why this is not the correct way to think about the situation is because initially there are three gold coins that could be chosen, and two out of the three of them are in a box with another gold coin.

For this reason, this puzzle feels quite similar to the more well-known Monty Hall problem, which is something you could possibly move on to. One advantage of the Monty Hall problem is that it is now so well known that there are simulators online that enable you to run a huge number of trials very quickly to verify the result. There are also lots of videos available that explain the solution.

Glossary

Mutually exclusive events

Sets of events or outcomes that are related such that, if one of them happens, then it means that none of the others can happen at the same time.

Example

When rolling a die, the outcomes are mutually exclusive because when one number lands on top it means that none of the others can.

Probability

The probability of an outcome (or event) is a measure of how likely that outcome is.

A value of 0 means that it is impossible, whereas 1 means that it is certain; otherwise the value must lie between 0 and 1. It may be given as a common fraction, a decimal fraction or a percentage.

Relative frequency

This is the experimental probability of an outcome, which is the value found after an activity has been repeated several times.
It is given by:

$$\frac{\text{number of times that named outcome(s) did happen}}{\text{number of times activity was done}}$$

Examples

A drawing-pin might be dropped several times and a count made of whether it landed 'point up' or 'point down'. Or a biased die might be rolled many times to determine the experimental probability of it landing on 5.

Sample space diagram

A sample space diagram is a table showing all the possible outcomes of an event.

Example

If two dice are rolled and the two numbers are added together, the possible outcomes can be displayed in a table like this:

		Dice 1					
		1	**2**	**3**	**4**	**5**	**6**
Dice 2	**1**	2	3	4	5	6	7
	2	3	4	5	6	7	8
	3	4	5	6	7	8	9
	4	5	6	7	8	9	10
	5	6	7	8	9	10	11
	6	7	8	9	10	11	12

From this table, it is clear that the most common total is 7.

Tree diagram

A diagram drawn to find and display all possible results when several outcomes are being combined.

Example

When 3 coins are flipped, all of the possible results can be found and displayed using a tree diagram like the one shown
(H = head, T = tail).

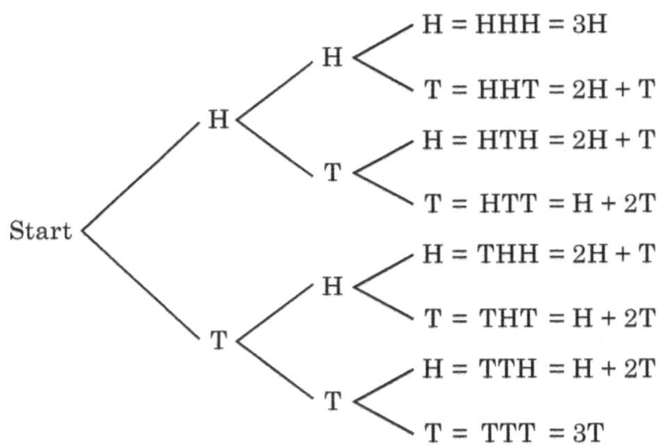

Venn diagram

A diagram used to give a pictorial view of the relationships between sets and subsets within a universal set; the universal set is shown enclosed by a rectangle, and all other sets are shown as circles or simple closed curves.

Examples

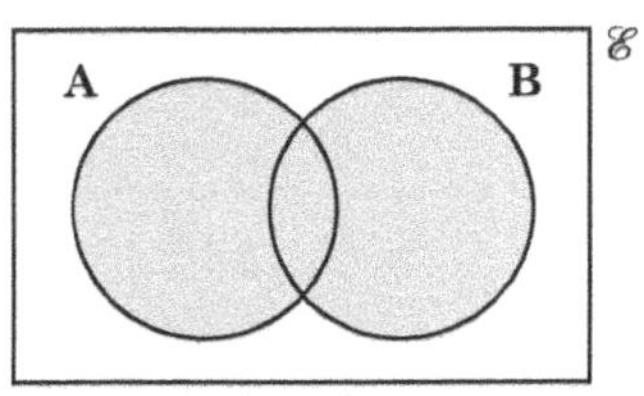

A ∪ B
union of A and B

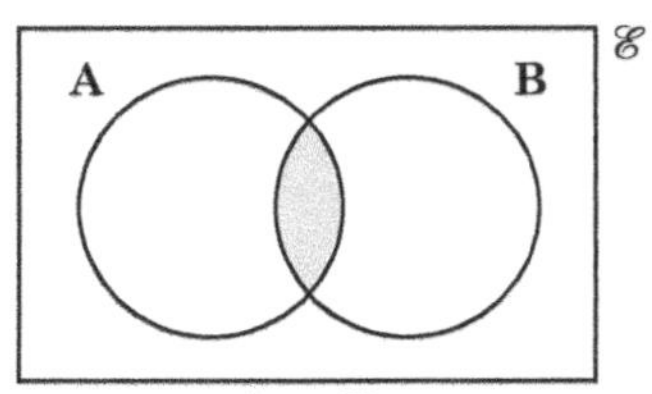

A ∩ B
intersection of A and B

Extension worksheet

This worksheet develops the work on tree diagrams to include three outcomes.

A bag contains ten coloured balls. Three of the balls are red, two of the balls are blue and the rest of the balls are yellow. One of the balls is removed from the bag, its colour is noted, and then it is replaced. A second ball is then removed from the bag.

Draw a tree diagram to show this situation and hence find the probability that the two balls removed from the bag are both the same colour.

To draw the tree diagram, you need three branches on each level:

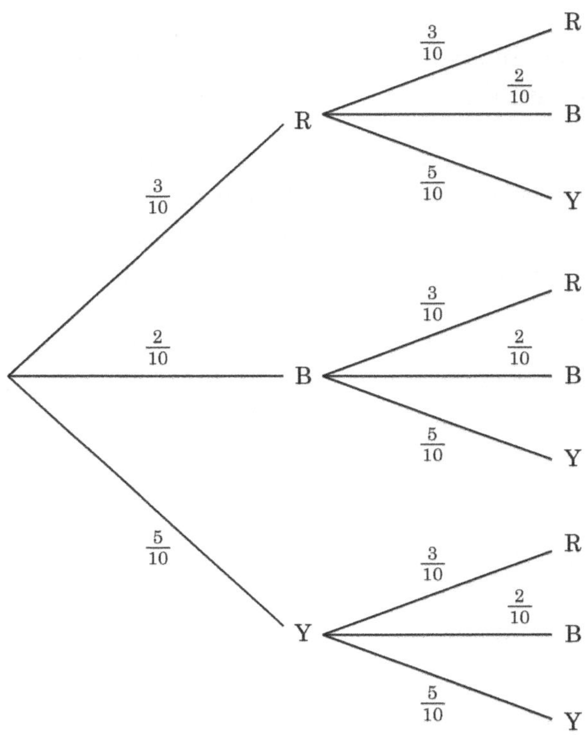

To find the probability of two red balls:

$$\frac{3}{10} \times \frac{3}{10} = \frac{9}{100}$$

To find the probability of two blue balls:

$$\frac{2}{10} \times \frac{2}{10} = \frac{4}{100}$$

To find the probability of two yellow balls:

$$\frac{5}{10} \times \frac{5}{10} = \frac{25}{100}$$

So, the probability that the two balls are the same colour is

$$\frac{9}{100} + \frac{4}{100} + \frac{25}{100} = \frac{38}{100} = \frac{19}{50}$$

1. On a stretch of road, there are two sets of traffic lights.

   For both sets, the probability that they are red when a driver arrives is 0.4, the probability that they are amber is 0.1 and the probability that they are green is 0.5.

   By drawing a tree diagram, or otherwise, find the probability that:

   a) both sets of lights are red

   b) both sets of lights are the same colour.

2. A bag contains 20 identical counters. Seven of them are coloured red, eight of them are coloured blue and the rest are coloured green. One counter is removed from the bag, its colour is noted and then it is replaced. A second counter is then removed from the bag. By drawing a tree diagram, or otherwise, find the probability that the two counters are different colours.

3. A bag contains 15 marbles. Six are green, five are red and four are yellow. Two marbles are removed from the bag, one after the other, and the first one is *not* replaced. By drawing a tree diagram, or otherwise, find the probability that the two marbles are the same colour.

(Answers on page 125)

10 Algebra 3

The main topic covered in this chapter is simultaneous equations and problems that can be solved using them.

The chapter starts, however, with an exercise on forming and solving equations.

It then continues with an exercise on substituting into formulae.

Changing the subject of a formula is also covered.

All these skills are required when the students come to forming and solving simultaneous equations.

Three methods of solution are covered:

- graphical solutions
- the elimination method
- the substitution method.

Prerequisite knowledge

This chapter assumes that students can solve linear equations and work with fractions.

Investigations, activities and puzzles

Simultaneous shapes

In the diagram, each shape is worth a particular value.

The total for each row is given.

What is the pentagon worth?

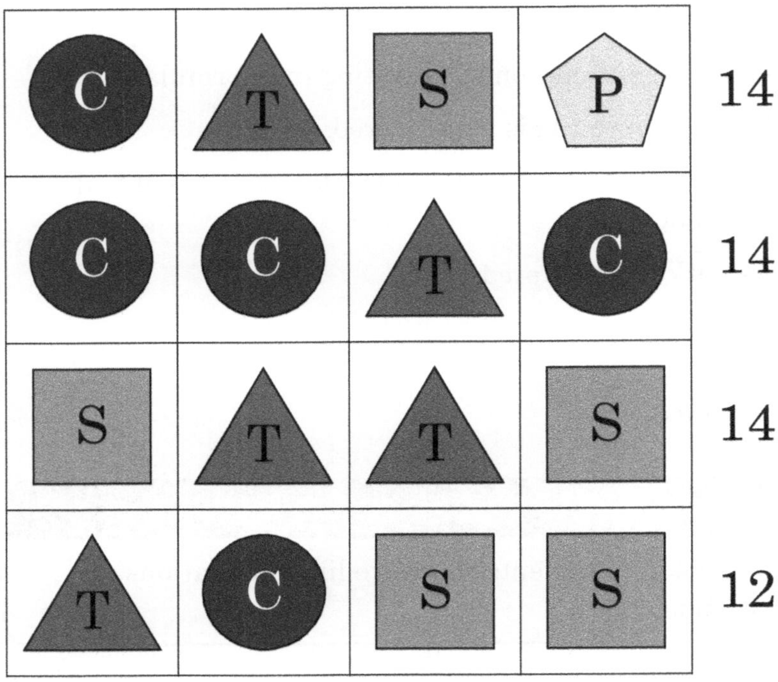

Notes on this activity

From rows 2, 3 and 4 we get three equations:

$$3C + T = 14, \text{ which means that } T = 14 - 3C$$

$$2S + 2T = 14, \text{ which means that } S + T = 7$$

$$2S + T + C = 12$$

Equation 1 tells us that $T = 14 - 3C$

Substituting the information from equation 2 into equation 3 tells us that

$$2(7 - T) + T + C = 12$$

Substituting the information from equation 1 into this equation tells us that

$$2(7 - (14 - 3C)) + (14 - 3C) + C = 12$$

Simplifying: $2(-7 + 3C) + 14 - 2C - 12 = 0$

$$-14 + 6C + 14 - 2C - 12 = 0$$

$$4C - 12 = 0$$

$$C = 3$$

Substituting back into equation 1:

$$3 \times 3 + T = 14, \text{ so } T = 5$$

Substituting back into equation 3:

$$2S + 5 + 3 = 12$$

$$2S = 4$$

$$S = 2$$

Looking now at the top row:

$$C + T + S + P = 14$$

So $3 + 5 + 2 + P = 14$

which means $10 + P = 14$

and $P = 4$

So the pentagon is worth 4.

These kinds of puzzles are very popular as they provide students with a fun application of simultaneous equations.

Here are some similar activities that you could consider with your students.

- Can they make their own puzzles of a similar nature?
- What is the maximum number of different shapes they can use in a given sized grid for the puzzle to be solvable?
- How does this relate to the solvability of equations with several variables?

Areas of obtuse triangles

For many triangles, the reason why the area is $\frac{1}{2} \times \text{base} \times \text{height}$ is easy to understand, because the triangle can be thought of as being half of a rectangle.

For example, it is easy to see here that the triangle is half the size of the rectangle in which it is contained.

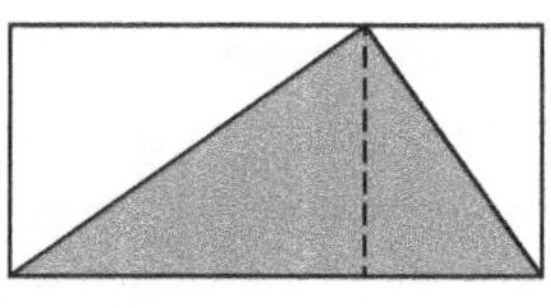

However, when the triangle is obtuse it is not so obvious.

Is the triangle here still half of the rectangle in which it is contained?

Does the formula $\frac{1}{2} \times \text{base} \times \text{height}$ still work with obtuse triangles?

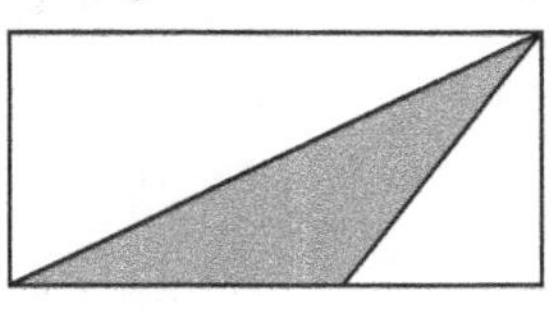

Notes on this activity

The formula does still work and this can be shown in two neat ways.

1. Although the obtuse triangle is not obviously half of a rectangle, the following diagram shows that it is half of a parallelogram.

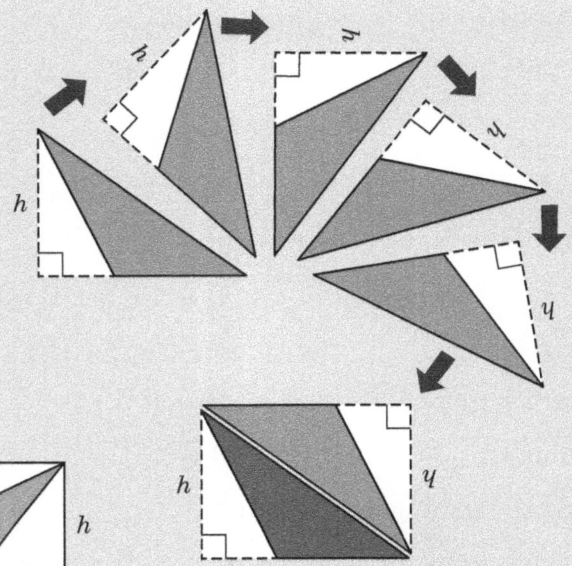

Since the formula for the area of a parallelogram is the same as the formula for the area of a rectangle, you can see that the formula for the area of a triangle area will still be applicable.

2. The area of the obtuse triangle is

$$\frac{1}{2}(b+x)h - \frac{1}{2}xh = \frac{1}{2}bh + \frac{1}{2}xh - \frac{1}{2}xh$$
$$= \frac{1}{2}bh$$

as required.

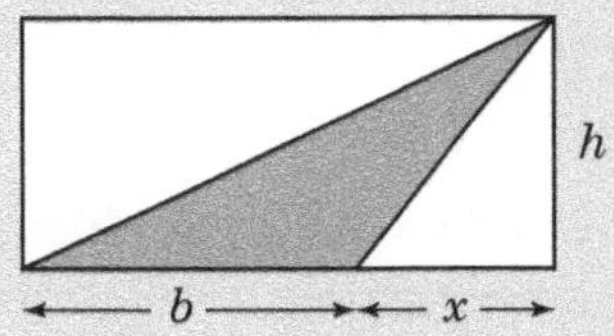

Glossary

Equation
An equation is a statement that indicates that two expressions (one of which may be a constant) have the same value.

Examples
$2x + 7 = 15$ and $3(x + 5) = 3x + 15$ are both equations.

Formula
A formula is a statement, written as an equation, that gives the relationship between two or more quantities, which are represented using symbols. When one or more of the values are known, the value of another quantity can be found.

Example
The volume V of a sphere can be found using the formula

$V = \pi d^3 \div 6$
if the value of the diameter d is known.

Simultaneous equations
A set of simultaneous equations consists of two (or more) equations whose variables all take the same value at the same time. Provided all the equations are independent, and a solution is possible, then n simultaneous equations containing n variables will have a unique solution. There are several ways of solving simultaneous equations, the most common being by a combination of substitution and elimination.

Substitution
A substitution in algebra is done by replacing one expression, or part of an expression, by something of equivalent value so that the overall truth of the original expression is unchanged.

Example
Given $4y + 3x = 22$ and $y = 2x$, then the second expression can be substituted for y in the first to give $4(2x) + 3x = 22$, which can be simplified to $11x = 22$.

Extension worksheet

This worksheet looks at two methods of solving simultaneous equations.

Sometimes you may want to solve a pair of simultaneous equations where one of the functions is a curve. In these situations, you will usually use substitution.

For example, if the two equations are $3x - 2y = 4$ and $x^2 - 4 = 0$, the second equation involves a quadratic curve.

There are two different ways to solve a problem like this. The first is to rearrange the linear equation and substitute it into the quadratic equation. The second is to solve the quadratic equation first, then substitute those results into the linear equation.

Method 1

First, rearrange the linear equation to make x the subject.

This gives $x = \frac{4 + 2y}{3}$

Substituting that into the second equation gives $\left(\frac{4 + 2y}{3}\right)^2 - 4 = 0$

Expanding the brackets gives $\frac{16 + 16y + 4y^2}{9} - 4 = 0$

Multiplying through by 9 to clear the fraction gives $16 + 16y + 4y^2 - 36 = 0$

Collecting terms and rearranging gives $4y^2 + 16y - 20 = 0$

Notice that each term can be divided by 4 to give $y^2 + 4y - 5 = 0$

This is now a quadratic equation. It can be factorised by reversing the process for expanding brackets.

Look for two numbers that multiply to make –5 and add to make 4.

These are 5 and –1, so you can factorise the equation as $(y + 5)(y - 1) = 0$

The product of these two brackets will be zero if either of the brackets equals zero.

This will be the case if $y = -5$ or $y = 1$.

Using the first equation, $x = \frac{4 + 2y}{3}$

When $y = -5$, $x = \frac{4 + 2(-5)}{3} = -2$. When $y = 1$, $x = \frac{4 + 2 \times 1}{3} = 2$

These are the solutions to the simultaneous equations.

Method 2

First, solve the quadratic equation.

The equation $x^2 - 4 = 0$ can be factorised as $(x + 2)(x - 2) = 0$.

You can check this by expanding the brackets.

This means that there are two solutions to the quadratic equation $x^2 - 4 = 0$.

Either $x = -2$ or $x = 2$.

These are the values of x that make one of the brackets equal to zero.

These two x values can now be substituted back into the linear equation $3x - 2y = 4$.

This can be rearranged to give $y = \dfrac{3x - 4}{2}$

If $x = -2$, then $y = \dfrac{3(-2) - 4}{2} = -5$

If $x = 2$, then $y = \dfrac{3 \times 2 - 4}{2} = 1$

Notice that these are the same results as obtained using method 1, but the amount of work required here is considerably less.

In the questions that follow, the line will always cut the curve in two places, which means there are two pairs of solutions, as in the example.

Try to solve some of the question using method 1 and some using method 2. Which method do you prefer?

Solve the following pairs of simultaneous equations.

1. $2x + 4y = 6$

$x^2 - 9 = 0$

2. $5x - 3y = 6$

$x^2 - 9 = 0$

3. $4x + 2y = 8$

$x^2 - 1 = 0$

4. $2x - y = 2$

$x^2 - 9 = 0$

5. $3x + 2y = 15$

$x^2 - 25 = 0$

6. $x + 2y = 2$

$x^2 - 4 = 0$

(Answers on page 126)

11 Shape and space 3

This chapter is made up of two halves. The first half deals with transformations and the second half deals with triangles.

The transformations covered are:

- reflection
- rotation
  - including finding the centre of rotation
- enlargement
  - including fractional scale factors
- translation
  - including the use of vector notation.

The work on triangles covers:

- Pythagoras' theorem
- similar triangles
- trigonometry in right-angled triangles
- problems involving trigonometry.

Prerequisite knowledge

This chapter assumes that students have a basic understanding of:

- the Cartesian coordinate system
- the equation of a straight line
- fractions
- angles.

Investigations, activities and puzzles

Pythagorean triples

A right-angled triangle can be drawn with integer (whole number) side lengths 3, 4 and 5 units.

This is called a '3 4 5 triangle'.

This works because $3^2 + 4^2 = 5^2$ from Pythagoras' theorem.

A 3-4-5 triangle is known as a Pythagorean triple.

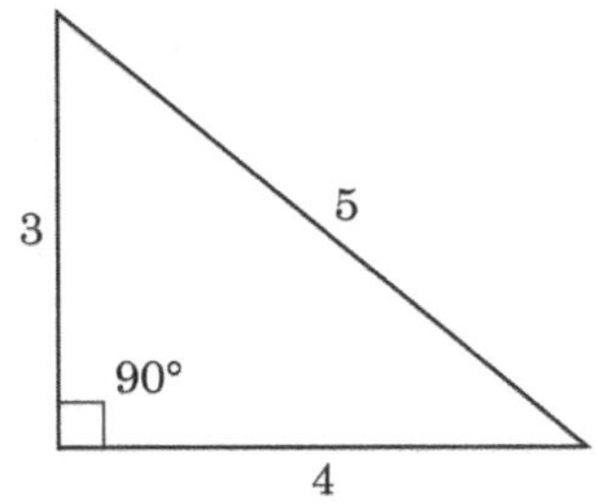

1. Can you find any other Pythagorean triples?
2. How many Pythagorean triples (using only whole numbers) exist?

Notes on this activity

Students will probably find 6, 8, 10 and 5, 12, 13 fairly quickly.

Discuss the fact that 6, 8, 10 is just an enlargement of 3, 4, 5, but 5, 12, 13 is not.

Clearly, because 6, 8, 10 is an enlargement of 3, 4, 5, this means that there are an infinite number of Pythagorean triples. Are there, however, an infinite number of triples that are not just multiples of each other?

Extension

1. Introduce the idea that Pythagorean triples can be generated by choosing two whole numbers, m and n, and then finding $m^2 + n^2$, $m^2 - n^2$, and $2mn$.

   Here are some questions that students could consider:

   - Which values of m and n give 3, 4, 5?
   - Which values of m and n give 5, 12, 13?
   - Use this method to generate other triples. What do you notice?
   - Can you show algebraically why this method works?
   - Do you think that every Pythagorean triple can be generated in this way? Why?

2. Tell the class about Fermat's last theorem. What can they find out about it?

Measuring the heights of trees

You can use trigonometry to measure the height of a tree or a tall building.

Stand looking up at the top of the tree and measure the angle of elevation through which you are looking. You can do this using a clinometer or a protractor.

Then use the tangent ratio to work out the height of the tree.

What will be true if the angle you are looking through is 45°?

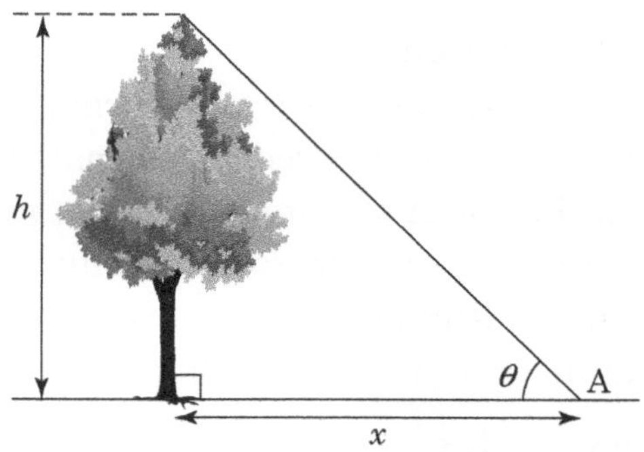

Notes on this activity

For example, if you are 10 metres away from the tree and the angle you measure is 50°, the calculation becomes $\tan 50° = \frac{h}{10}$, so $h = 10\tan 50° = 11.92$ m

Then you need to add the height above the floor at which you are holding your angle measuring device. If you do not have a clinometer you can use a square piece of paper and cut along its diagonal. This will give you a 45° angle. If you walk away from the tree, while looking up through a 45° angle until the paper lines up with the top of the tree, the distance you are away from the tree will be the same as the height of the tree.

Enlarging a picture

Find a picture online or from a magazine and enlarge it.

- Draw rays from the centre of enlargement through several points on the original picture.
- Measure the distance from the centre of enlargement to each point along the rays.
- Multiply these distances by a chosen scale factor.
- Measuring from the centre of enlargement, mark the new positions of the points on the rays.
- Join the new points together to make the enlarged picture.

Notes on this activity

Students need to carefully consider what their scale factor of enlargement is going to be, based on the size of the paper they are working on. They also need to carefully consider where they are going to put their centre of enlargement, as it will determine how the two pictures are situated relative to each other.

Glossary

Enlargement

An enlargement is a transformation in which the distances between every pair of points on the object are multiplied by the same value (the scale factor) to produce the image. The multiplier can take any value – a whole number or a fraction – but not zero.

Examples

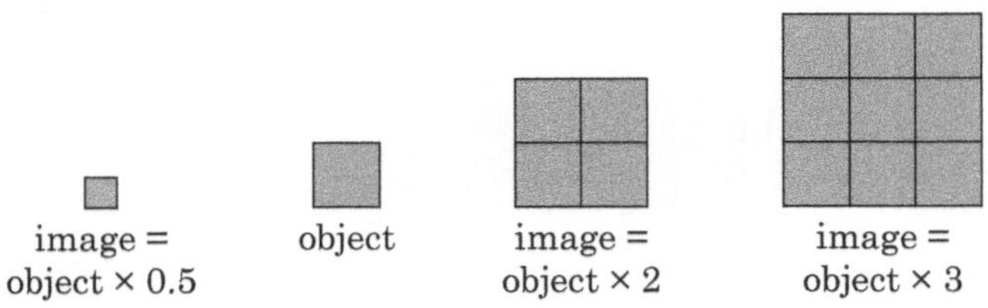

Hypotenuse

The hypotenuse is the side of a right-angled triangle that is opposite the right angle. It is the longest side of the triangle.

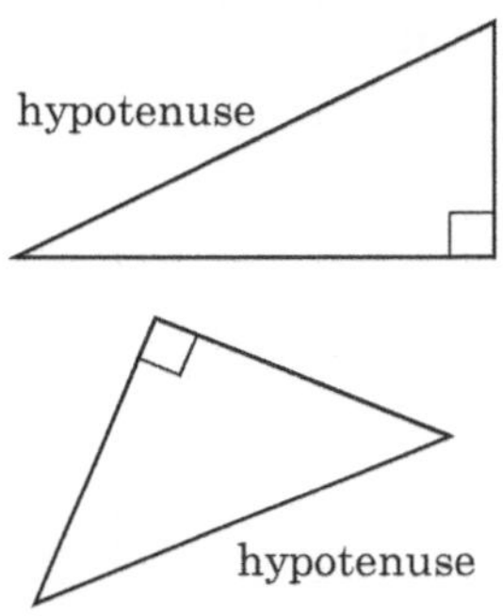

Pythagoras' theorem

This theorem states that, in any right-angled triangle, the square of the hypotenuse is equal to the sum of the squares of the other two sides.

Algebraically, this means that, for any right-angled triangle,

$$a^2 + b^2 = c^2$$

where c is the hypotenuse (the side opposite the right angle) and a and b are the other two sides.

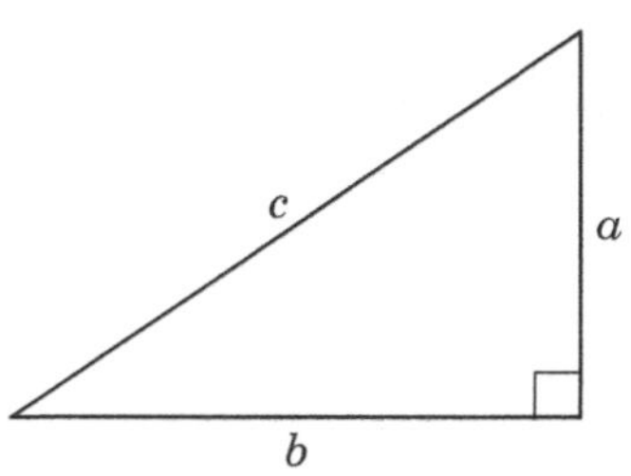

Rotation

A rotation is a transformation about a fixed point such that every point in the object turns through the same angle relative to that fixed point. A rotation is described by giving the angle and direction of the turn, and the position of the fixed point about which the turn is made.

Examples

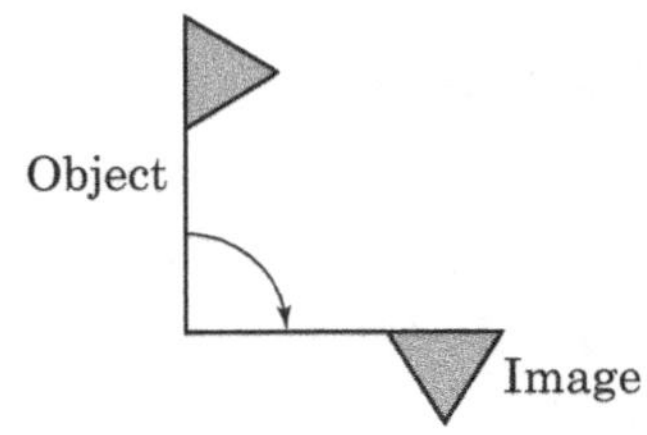

A rotation of one-quarter of a turn clockwise about a point on the bottom of the object

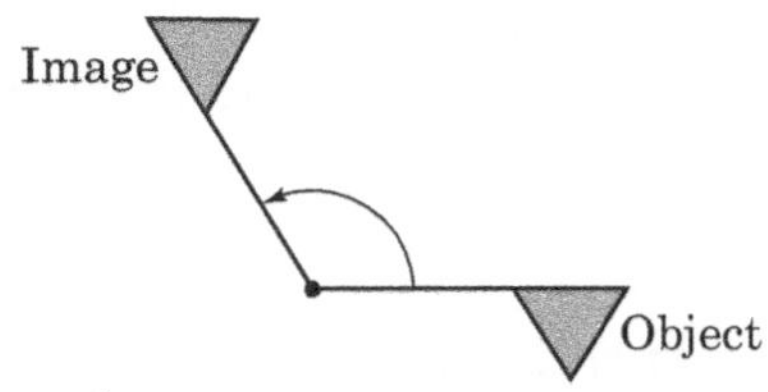

A rotation of 120° anticlockwise about the fixed point marked

Similar

Geometrical figures are similar if they have the same angles but a different size, and their lengths are in the same proportion. One shape is an enlargement of the other. Corresponding (matching) angles in each shape will be the same size.

Examples

All of these triangles are similar:

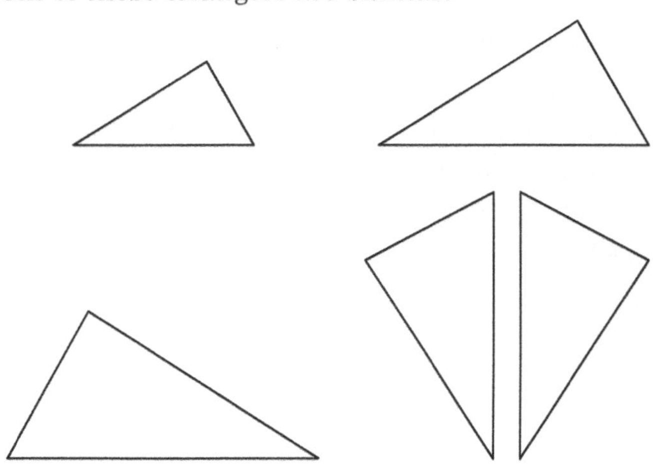

Translation

A translation is a transformation in which every point in the object is moved to a new position by the same combination of horizontal and vertical movements. A translation is described by the direction and length of each movement. The size and orientation of an object do not change when it is translated.

Examples

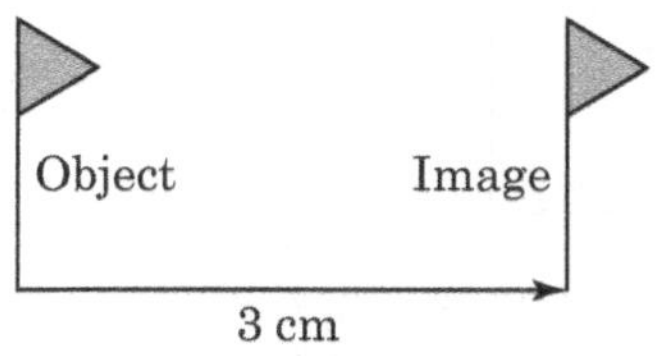

A translation of 3 cm right

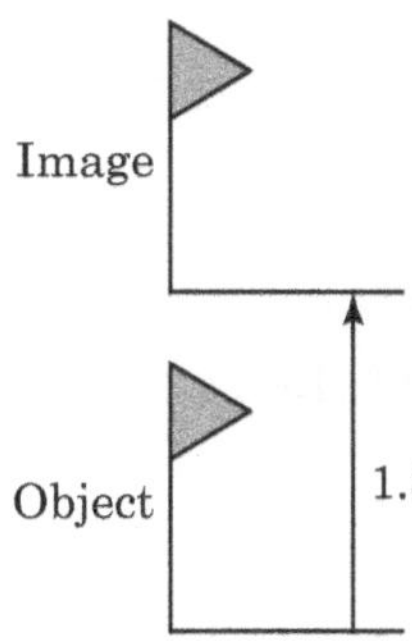

A translation of 1.5 cm up

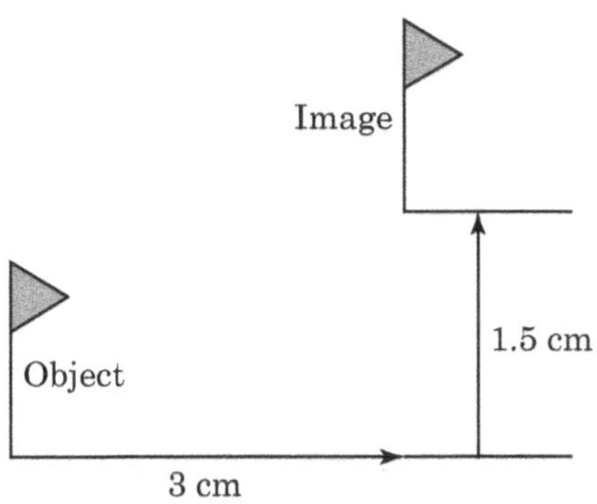

A translation of 3 cm right and 1.5 cm up

Trigonometry

This is the study of triangles with regard to measurements of their angles and side lengths. It uses the trigonometric ratios sine, cosine and tangent to describe the relationships between these measurements and calculate unknown values.

Extension worksheet

This worksheet extends the work on Pythagoras' theorem into three dimensions.

Pythagoras' theorem states that in a right-angled triangle, the sum of the squares of the shorter two sides is equal to the square of the hypotenuse. So:

$$a^2 + b^2 = c^2$$

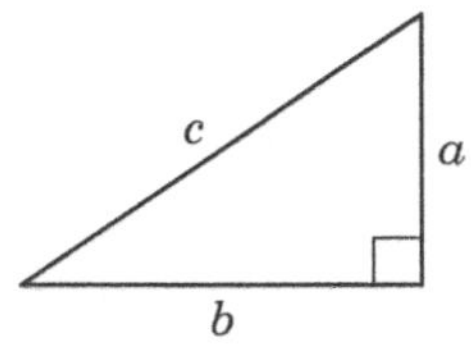

The theorem can be extended to three dimensions and used to find the length of the long diagonal in a cuboid.

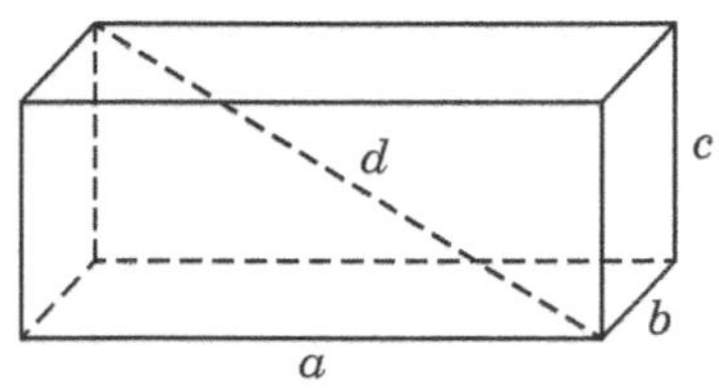

Here, the sum of the squares of the three sides of the cuboid is equal to the square of the long diagonal. So:

$$a^2 + b^2 + c^2 = d^2$$

Use this new formula to work out the length of the long diagonal, d, for the following cuboids.

1. $a = 6\,\text{cm}$, $b = 4\,\text{cm}$ and $c = 3\,\text{cm}$
2. $a = 7\,\text{cm}$, $b = 3\,\text{cm}$ and $c = 2\,\text{cm}$
3. $a = 10\,\text{cm}$, $b = 5\,\text{cm}$ and $c = 6.5\,\text{cm}$
4. $a = 25\,\text{cm}$, $b = 12.2\,\text{cm}$ and $c = 11\,\text{cm}$
5. $a = 8.6\,\text{cm}$, $b = 4.8\,\text{cm}$ and $c = 2.3\,\text{cm}$

Extension

6. Write down a suitable variation of the above formula that enables you to calculate the length of a when given the lengths of b, c and d.
7. Use this new formula to calculate the length of side a for the cuboid for which $b = 7\,\text{cm}$, $c = 6\,\text{cm}$ and $d = 18\,\text{cm}$.

(Answers on page 126)

Revision checklists

Number

Make sure you can...

	Revised	Secure
• add, subtract, multiply and divide whole numbers, decimals and fractions		
• recognise: • natural numbers • integers • prime numbers • square numbers • cube numbers • common factors • common multiples • rational numbers • irrational numbers • reciprocals		
• continue a given number sequence		
• find the nth term for a given number sequence		
• make estimates and round answers to an appropriate degree of accuracy		
• work out upper and lower bounds for a given quantity		
• use standard form		
• understand ratio, including map scales		
• understand direct proportion, including currency conversion		
• calculate percentage increase and decrease		
• solve problems involving: • simple interest • compound interest		
• use a calculator efficiently		
• convert between fractions, decimals and percentages		

Algebra and graphs

Make sure you can…

	Revised	Secure
• substitute numbers into expressions and formulae		
• construct and rearrange formulae		
• expand brackets		
• factorise expressions		
• solve linear equations in one unknown		
• solve practical problems using algebra		
• solve pairs of linear simultaneous equations		
• solve practical problems involving linear simultaneous equations		
• understand and use the rules of indices		
• represent and interpret inequalities, including using number lines		
• recognise, sketch and interpret graphs of these functions: • linear • quadratic		
• plot graphs of straight lines and curves		
• solve equations graphically		
• plot and interpret distance–time and speed–time graphs		

Coordinate geometry

Make sure you can…

	Revised	Secure
• obtain the equation of a straight line in the form $y = mx + c$		
• calculate the gradient of a straight line given two coordinates		
• find the gradient of a parallel line		
• find the equation of a parallel line		

Geometry

Make sure you can…

	Revised	Secure
• measure and draw lines and angles		
• construct triangles given three sides		
• find missing angles in: • triangles • quadrilaterals • irregular polygons • regular polygons		
• know and use rules for angles: • at a point • on a straight line • in parallel lines		
• know and use circle theorems		
• recognise and describe rotational symmetry		
• recognise and describe line symmetry		
• understand and describe what is meant by similarity		
• calculate missing lengths in similar shapes		
• draw and recognise nets of solids		
• use and draw three-figure bearings		
• apply scale drawings to solve problems		

Mensuration

Make sure you can...

	Revised	Secure
• calculate the perimeter of a rectangle		
• calculate the area of a rectangle		
• calculate the perimeter of a triangle		
• calculate the area of a triangle		
• calculate the perimeter of a parallelogram		
• calculate the area of a parallelogram		
• calculate the perimeter of a trapezium		
• calculate the area of a trapezium		
• calculate the area of a circle		
• calculate the circumference of a circle		
• calculate arc length		
• calculate sector area		
• calculate the surface area and volume of: • a cone • a pyramid • a prism • a cylinder • a sphere		
• convert between units including units of area and volume		
• calculate the area of compound shapes		
• calculate the volume of compound solids		

Trigonometry

Make sure you can...

	Revised	Secure
• use Pythagoras' theorem		
• apply the sine, cosine and tangent ratios to find unknown sides		
• apply inverse sine, cosine and tangent to find unknown angles		
• solve problems in two dimensions using Pythagoras' theorem and trigonometry		

Transformations and vectors

Make sure you can...

	Revised	Secure
• apply the following transformations to a shape: • rotation • reflection • translation • enlargement • use vector notation to describe a translation		

Probability

Make sure you can…

	Revised	Secure
• describe sets using appropriate notation and language		
• draw Venn diagrams		
• calculate simple probabilities		
• understand how to calculate the probability of an event *not* happening		
• calculate relative frequency		
• calculate expected frequency		
• calculate the probability of combined events using: • sample space diagrams • tables • tree diagrams • Venn diagrams		

Statistics

Make sure you can...

	Revised	Secure
• construct the following types of chart: • bar chart • pie chart • pictogram • simple frequency distribution • stem-and-leaf diagram		
• calculate the mean, median, mode and range from discrete data		
• compare data sets using diagrams and calculations		
• draw and interpret a scatter diagram		
• understand what is meant by positive, negative and no correlation		
• draw and interpret a line of best fit		

Multiple choice revision test 1

1. In standard form the value of $2000 \times 80\,000$ is:

A 16×10^6 **B** 1.6×10^9

C 1.6×10^7 **D** 1.6×10^8

2. In the triangle the size of angle x is:

A 35°

B 70°

C 110°

D 40°

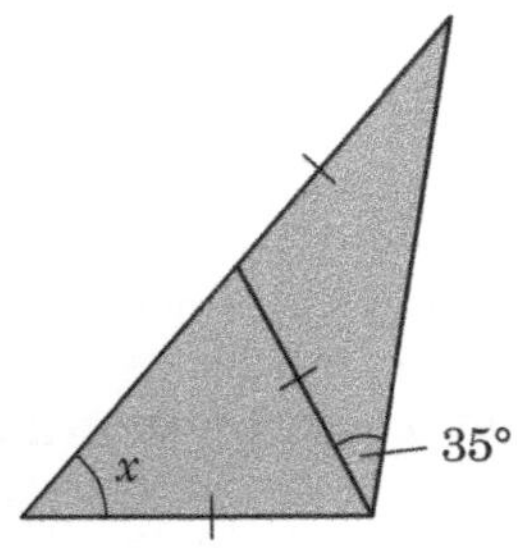

3. The approximate value of $\dfrac{3.96 \times (0.5)^2}{97.1}$ is:

A 0.01 **B** 0.02

C 0.04 **D** 0.1

4. Which of the following is/are correct?

1. $\sqrt{0.16} = 0.4$ **2.** $0.2 \div 0.1 = 0.2$ **3.** $\dfrac{4}{7} > \dfrac{3}{5}$

A 1 only **B** 2 only

C 3 only **D** 1 and 2

5. How many prime numbers are there between 30 and 40?

A 0 **B** 1

C 2 **D** 3

6. $\dfrac{1}{3} + \dfrac{2}{5} = ?$

A $\dfrac{2}{8}$ **B** $\dfrac{3}{8}$

C $\dfrac{3}{15}$ **D** $\dfrac{11}{15}$

7. In the triangle the size of the largest angle is:

A 30°

B 90°

C 120°

D 80°

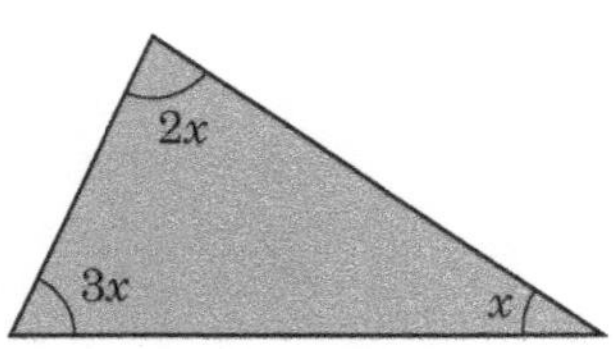

8. Given $a = \frac{3}{5}$, $b = \frac{1}{3}$, $c = \frac{1}{2}$ then:

A $a < b < c$ **B** $a < c < b$

C $a > b > c$ **D** $a > c > b$

9. What is the value of $1 - 0.05$ as a fraction?

A $\frac{1}{20}$ **B** $\frac{9}{10}$

C $\frac{19}{20}$ **D** $\frac{5}{100}$

10. Adjacent angles in a parallelogram are $x°$ and $3x°$.
The smallest angles in the parallelogram are each:

A 30° **B** 45°

C 60° **D** 120°

11. The bearing of A from B is 120°.
What is the bearing of B from A?

A 060° **B** 120°

C 240° **D** 300°

12. The number of letters in the word SNAIL that have line symmetry is:

A 0 **B** 1

C 2 **D** 3

(Answers on page 127)

Multiple choice revision test 2

1. The mean weight of a group of 11 men is 70 kg. What is the mean weight of the remaining group when a man of weight 90 kg leaves?

   A 80 kg **B** 72 kg

   C 68 kg **D** 62 kg

2. A man paid tax on $9000 at 30%. He paid the tax in 12 equal payments. Each payment was:

   A $2.25 **B** $22.50

   C $225 **D** $250

3. A man is paid $180 per week *after* a pay rise of 20%. What was he paid before?

   A $144 **B** $150

   C $160 **D** $164

4. Given the equation $5^x = 120$, the best approximate value for x is:

   A 2 **B** 3

   C 4 **D** 25

5. A firm employs 1200 people, of whom 240 are men. The percentage of employees who are men is:

   A 40% **B** 10%

   C 15% **D** 20%

6. 800 decreased by 5% is:

   A 795 **B** 640

   C 760 **D** 400

7. $240 is shared in the ratio 2 : 3 : 7. The largest share is:

   A $130 **B** $140

   C $150 **D** $160

8. When the sides of the square are increased by 10% the area is increased by:

   A 10% **B** 20%

   C 21% **D** 15%

9. The price of a TV changed from \$240 to \$300.
What is the percentage increase?

A 15% **B** 20%

C 60% **D** 25%

10. A school has 400 students, of whom 250 are boys.
The ratio of boys to girls is:

A 5:3 **B** 3:2

C 3:5 **D** 8:5

11. The approximate value of $\frac{9.65 \times 0.203}{0.0198}$ is:

A 99 **B** 9.9

C 0.99 **D** 180

12. What is the new fare when the old fare of \$250 is increased by 8%?

A \$258 **B** \$260

C \$270 **D** \$281.25

13. The mean of four numbers is 12.
The mean of three of the numbers is 13.
What is the fourth number?

A 9 **B** 12.5

C 7 **D** 1

14. The value of 4865.355 correct to 2 significant figures is:

A 4865.36 **B** 4865.35

C 4900 **D** 49

15. Here are some numbers: 7, 3, 1, 4, 6
The median is:

A 1 **B** 4

C 4.2 **D** 3.5

(Answers on page 127)

Multiple choice revision test 3

1. How many mm are there in 1 m 1 cm?

A 1001 **B** 1110

C 1010 **D** 1100

2. The circumference of a circle is 16π cm.
The radius, in cm, of the circle is:

A 2 **B** 4

C $\frac{4}{\neq}$ **D** 8

3. The line $y = 2x - 1$ cuts the x-axis at P. The coordinates of P are:

A $(0, -1)$ **B** $\frac{1}{2},0$

C $-\frac{1}{2},0$ **D** $(-1, 0)$

4. How many square tiles of side 50 cm will be needed to cover the floor shown?

A 16

B 32

C 64

D 84

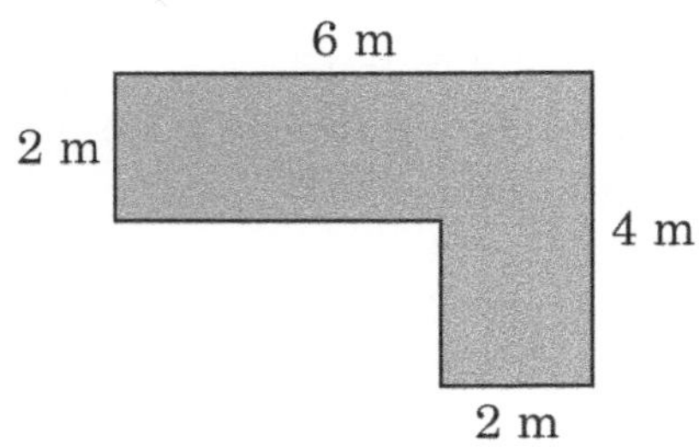

5. A car travels for 20 minutes at 45 km/h and then for 40 minutes at 60 km/h. The average speed for the whole journey is:

A 52 km/h **B** 50 km/h

C 54 km/h **D** 55 km/h

6. What is the area, in cm^2, of a circle of diameter 10 cm?

A 10π **B** 25π

C 49π **D** 100π

7. The volume, in cm^3, of the cylinder is:

A 9π

B 12π

C 600π

D 900π

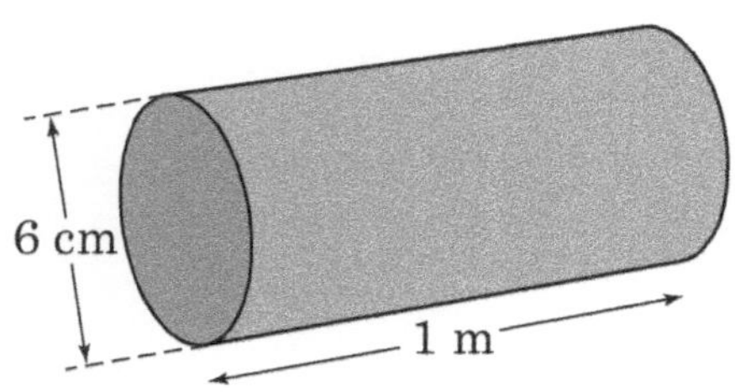

8. Two dice numbered 1 to 6 are thrown together and their scores are added.
The probability that the sum will be 12 is:

A $\frac{1}{6}$ **B** $\frac{1}{12}$

C $\frac{1}{36}$ **D** $\frac{1}{18}$

9. The length, in cm, of the minor arc is:

A 2π

B 3π

C 6π

D $13\frac{1}{2}\pi$

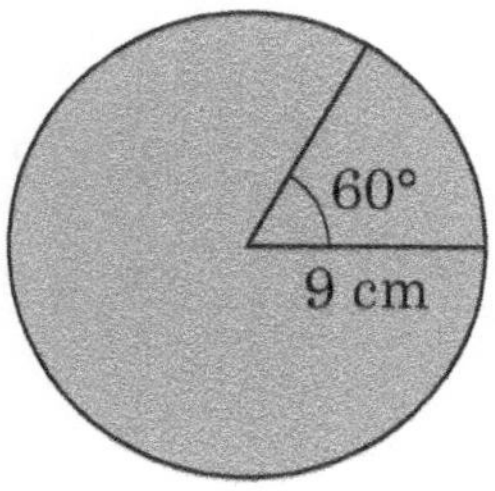

10. What is the area of this triangle?

A $12x^2$

B $15x^2$

C $16x^2$

D $30x^2$

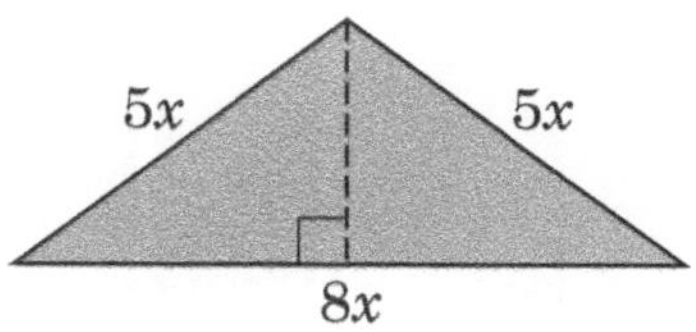

11. What is the area, in m^2, of a square with each side 0.02 m long?

A 0.0004 **B** 0.004

C 0.04 **D** 0.4

12. A fair coin is tossed three times.
The probability of getting three 'heads' is:

A $\frac{1}{3}$ **B** $\frac{1}{6}$

C $\frac{1}{8}$ **D** $\frac{1}{16}$

(Answers on page 127)

Multiple choice revision test 4

1. The formula $\frac{x}{a} + b = c$ is rearranged to make x the subject. What is x?

A $a(c - b)$ **B** $ac - b$

C $\frac{c - b}{a}$ **D** $ac = ab$

2. Given that $\frac{3}{n} = 5$, then $n = ?$

A 2 **B** −2

C $1\frac{2}{3}$ **D** 0.6

3. What is the sine of 45°?

A 1

B $\frac{1}{2}$

C $\frac{1}{\sqrt{2}}$

D $\sqrt{2}$

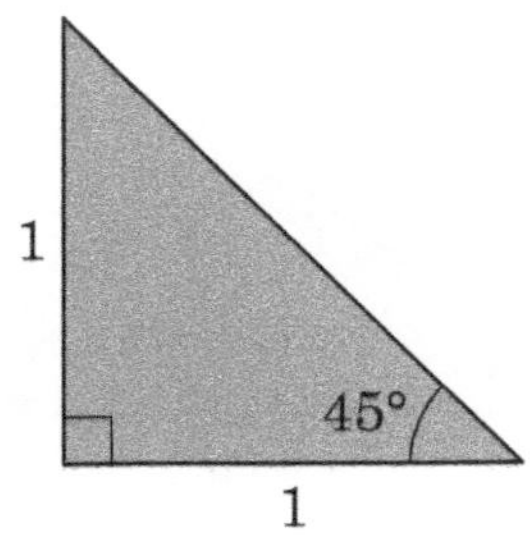

4. What is the value of the expression $(x - 2)(x + 4)$ when $x = -1$?

A 9 **B** −9

C 5 **D** −5

5. What are the coordinates of the point $(1, -1)$ after reflection in the line $y = x$?

A $(-1, 1)$ **B** $(1, 1)$

C $(-1, -1)$ **D** $(1, -1)$

6. Find the length x.

A 5

B 6

C 8

D $\sqrt{50}$

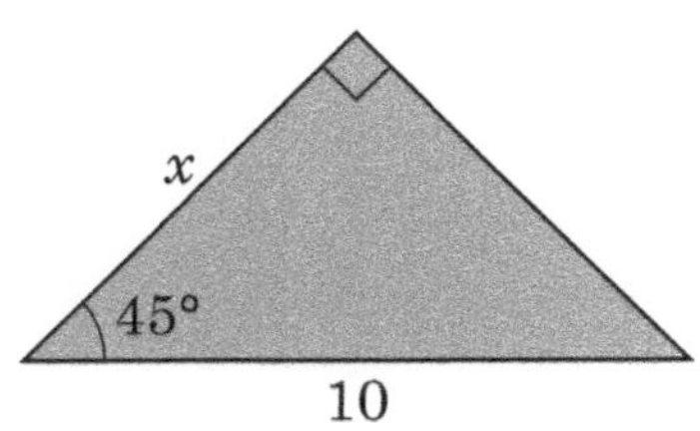

7. Given that $m = 2$ and $n = -3$, find the value of mn^2.

A -18 | B 18

C -36 | D 36

8. Find the length x.

A 6

B 5

C $\sqrt{44}$

D $\sqrt{18}$

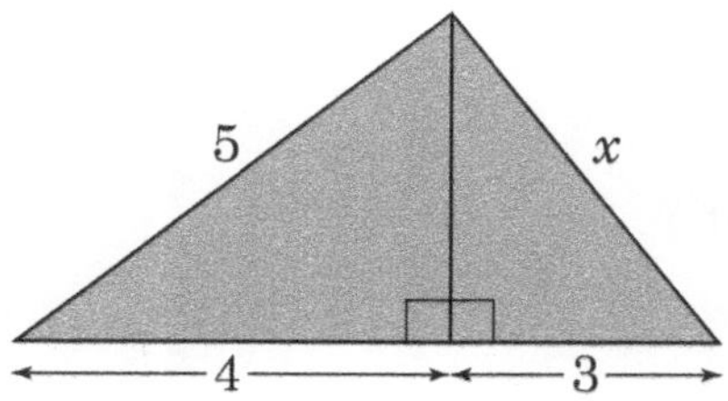

9. Numbers m, x and y satisfy the equation $y = mx^2$.
When $m = \frac{1}{2}$ and $x = 4$ the value of y is:

A 4 | B 8

C 1 | D 2

10. State which point does *not* lie on the curve $y = \frac{12}{x}$.

A $(6, 2)$ | B $\frac{1}{2}, 24$

C $(-3, -4)$ | D $(3, -4)$

11. I start with x, then square it, multiply by 2 and finally subtract 3.
The final result is:

A $(2x)^2 - 3$ | B $(2x - 3)^2$

C $2x^2 - 3$ | D $2(x - 3)^2$

12. What is the value of x satisfying these simultaneous equations:
$3x + 2y = 13$
$x - 2y = -1$?

A 7 | B 3

C $3\frac{1}{2}$ | D 2

13. State the coordinates of the point $(2, -2)$ after reflection in the line $y = -x$?

A $(-2, 2)$ | B $(2, -2)$

C $(-2, -2)$ | D $(2, 2)$

14. Solve the equation $2(x - 3) + 5 = 7$.

A $x = 1$ | B $x = 2$

C $x = 4$ | D $x = 8$

(Answers on page 127)

Examination-style Paper 1 (Non-calculator)

Paper 1 Non-calculator (Core)

1 hour 30 minutes

- You must answer **all** questions.
- Do **not** use a calculator.
- You will need geometrical instruments.
- All necessary working should be shown clearly.
- The total mark for this paper is 80.

List of formulae

Area, A, of triangle, base b, height h	$A = \frac{1}{2}bh$
Area, A, of circle of radius r	$A = \pi r^2$
Circumference, C, of circle of radius r	$C = 2\pi r$
Curved surface area, A, of cylinder of radius r, height h	$A = 2\pi rh$
Volume, V, of cylinder of radius r, height h	$V = \pi r^2 h$
Curved surface area, A, of cone of radius r, sloping edge l	$A = \pi rl$
Volume, V, of cone of radius r, height h	$V = \frac{1}{3}\pi r^2 h$
Surface area, A, of sphere of radius r	$A = 4\pi r^2$
Volume, V, of sphere of radius r	$V = \frac{4}{3}\pi r^3$
Volume, V, of prism, cross-sectional area A, length l	$V = Al$
Volume, V, of pyramid, base area A, height h	$V = \frac{1}{3}Ah$

Do **not** use calculators.

1. Marco swam a total of 1500 m.

He swam under water for 845 m and on the surface for the remaining distance.

a) Work out the distance he swam on the surface.

________ m **[1]**

b) Write the number 845 in words.

____

____ **[1]**

2. Philomena measures the length of her pencil as 240 mm to the nearest 10 mm.

State the lower bound and the upper bound for the length of Philomena's pencil.

Lower bound = ________ mm **[1]**

Upper bound = ________ mm **[1]**

3. $a = kb + c$

Make k the subject of the formula.

$k =$ ________ **[2]**

4. The diagram shows a net of a solid.

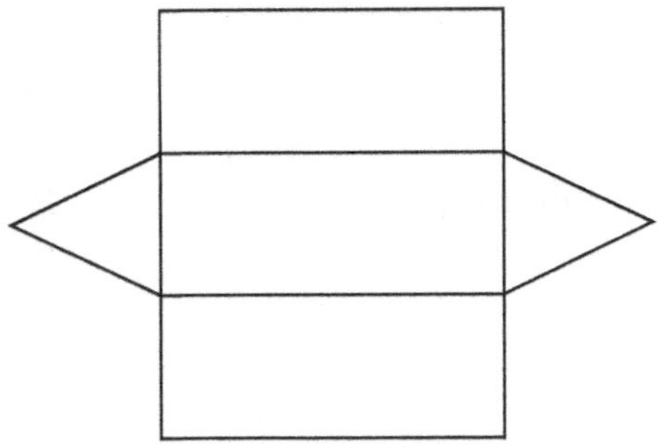

a) Write down the name of the solid.

_______________ **[1]**

b) For the solid, write down the number of edges.

_______________ **[1]**

5. Albert, Barry and Christine go to the supermarket.

The money they have is in the ratio

Albert : Barry : Christine = 4 : 6 : 7

Albert has $16.

a) How much money do they have in total?

$ _________ **[2]**

b) Barry spends $7 on a pizza.

How much money does he have left?

$ _________ **[1]**

6. Write down the number of square millimetres in one square centimetre.

_______________ **[1]**

7. Factorise completely

$3a^2 - 6ab$

______________________ **[2]**

8. Work out

a) $5 + 9 \times 4$

______________________ **[1]**

b) $-7 - (-11)$

______________________ **[1]**

c) 5^{-2}

______________________ **[2]**

9. **a)** Draw all the lines of symmetry on the shapes below. (Shape B is a regular polygon.)

Shape *A*:

Shape *B*:

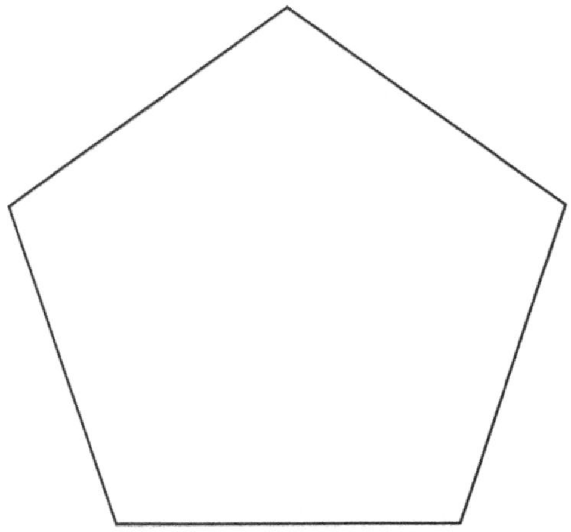

[2]

b) Write down the order of rotational symmetry of shape *B*.

______________ **[1]**

10. The diagram shows a shape made from two different triangles.

The shape has an area of $54\,\text{cm}^2$.

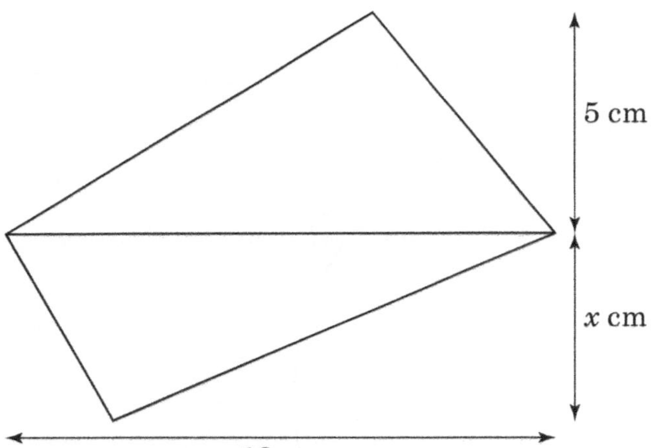

Find the value of x.

$x =$ ______________ **[4]**

11. Davide plots a scatter diagram of height against weight.

The graph shows that as height increases, weight increases.

Which of these types of correlation does the scatter diagram show?

negative correlation **positive correlation** **no correlation**

_______________________________ **[1]**

12.

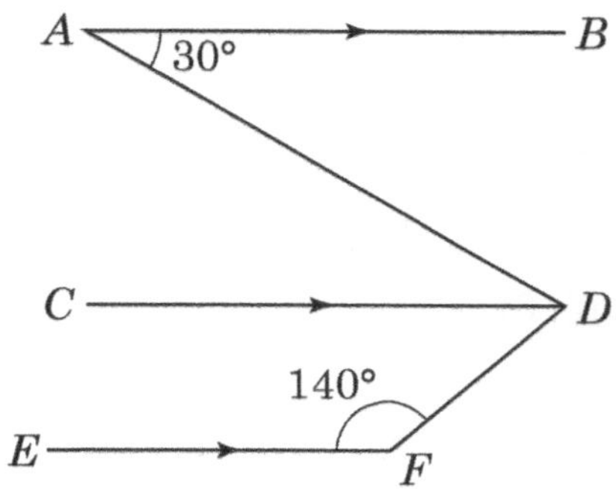

The diagram shows three parallel lines: *AB*, *CD* and *EF*.

Angle *BAD* is 30° and angle *DFE* is 140°.

Calculate angle *ADF*.

Angle *ADF* = _______________________________ **[2]**

13. Solve the equation $2 - 3x = 2x + 7$.

$x =$ ________________ **[2]**

14. A straight line, l, crosses the x-axis at (2, 0) and the y-axis at (0, 8).

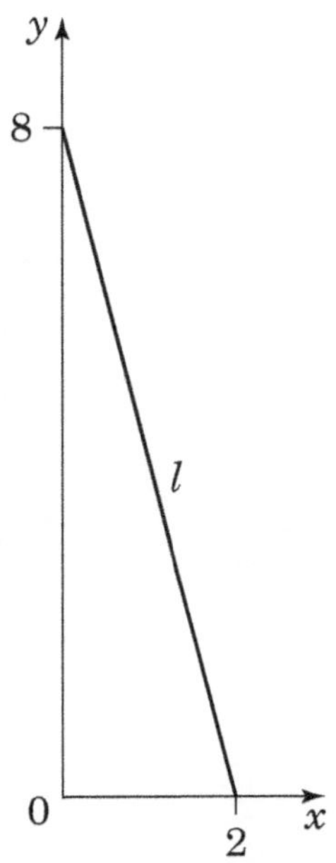

NOT TO SCALE

a) Find the gradient of line l.

________________ **[1]**

b) Write down the equation of the line l in the form $y = mx + c$.

$y =$ ________________ **[1]**

15. The diagrams show a sequence of shapes made from matchsticks.

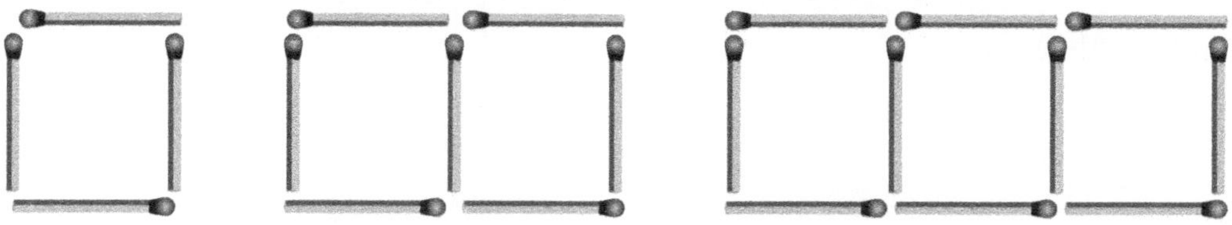

1 2 3

The table gives the number of matchsticks in each shape.

Shape number	1	2	3
Number of matchsticks	4	7	10

Find the number of matchsticks in

a) shape 4

______________ [1]

b) shape 10

______________ [1]

c) shape n.

______________ [2]

16. a) Write down the smallest **odd** integer that satisfies the inequality $n > 3.5$

$n =$ ______________ [1]

b) Write an inequality, in terms of x, to represent the interval shown on this number line.

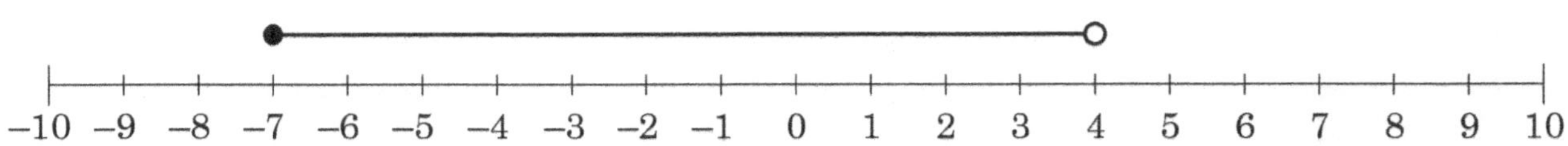

______________ [2]

17. Adrian spent \$30 on a new shirt.

This amount is $\frac{5}{6}$ of his allowance.

Calculate his allowance.

\$ ______________________ **[2]**

18. Expand the brackets and simplify $4x - 5(2x - 3)$.

______________________ **[2]**

19. Write down a multiple of 8 and 12 which is less than 40.

______________________ **[1]**

20. Solve the simultaneous equations.

$4x - 5y = 22$

$9x + 10y = 7$

$x =$ ______________________

$y =$ ______________________ **[3]**

21. a) Complete the table of values for $y = 2 + 3x - x^2$.

x	-3	-2	-1	0	1	2	3	4	5
y	-16	-8				4	2		-8

[3]

b) Draw the graph of $y = 2 + 3x - x^2$.

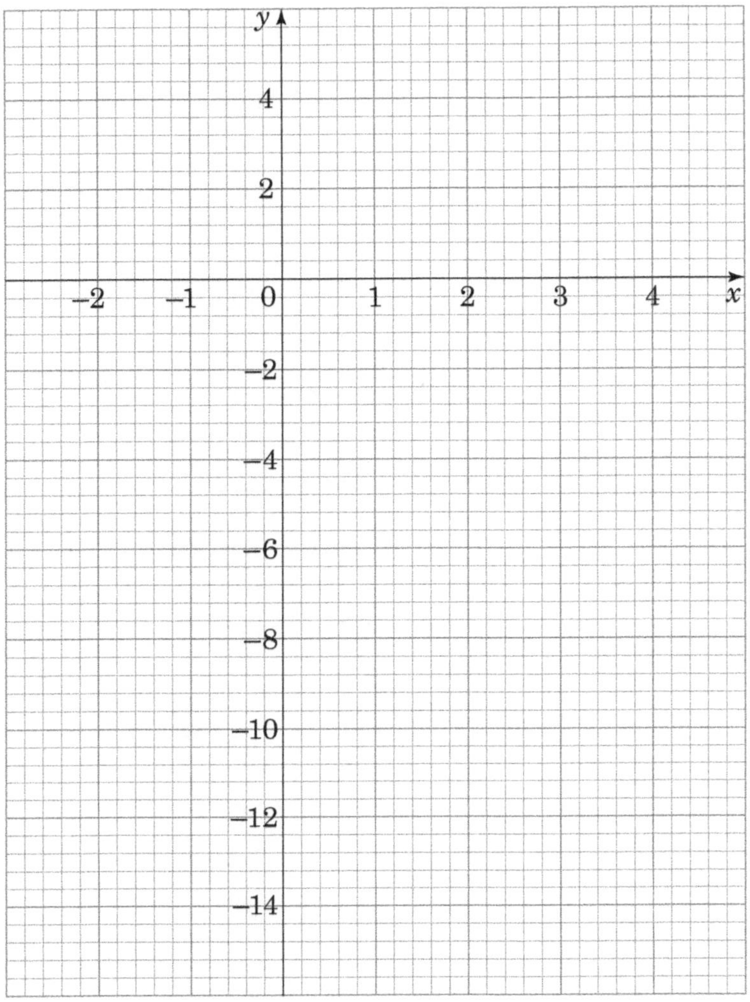

[4]

c) Use your graph to find the solutions to the equation $2 + 3x - x^2 = 0$.

$x =$ ____________________

or $x =$ ____________________ **[2]**

d) i) On the grid, draw the line of symmetry of the graph. **[1]**

ii) Write down the equation of the line of symmetry.

____________________ **[1]**

22. A bag contains 8 orange discs and 4 purple discs. A disc is selected from the bag at random, the colour is noted and then the disc is replaced.

A second disc is then selected.

a) Draw a tree diagram to represent this situation.

[2]

b) Find the probability that the two discs are the same colour.

_______________ **[2]**

23. *ABCDE* is a regular pentagon with centre *O*.

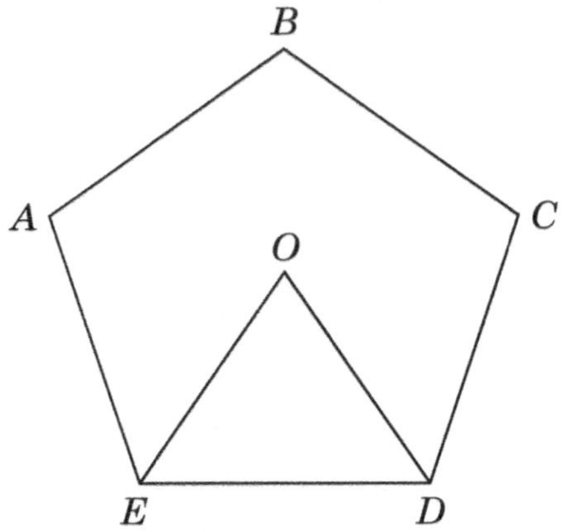

NOT TO SCALE

a) Calculate the size of angle *AED*.

____

____

______________ **[2]**

b) Calculate the size of angle *EOD*.

____

____

______________ **[2]**

c) Write down the size of angle *AEO*.

______________ **[1]**

24.

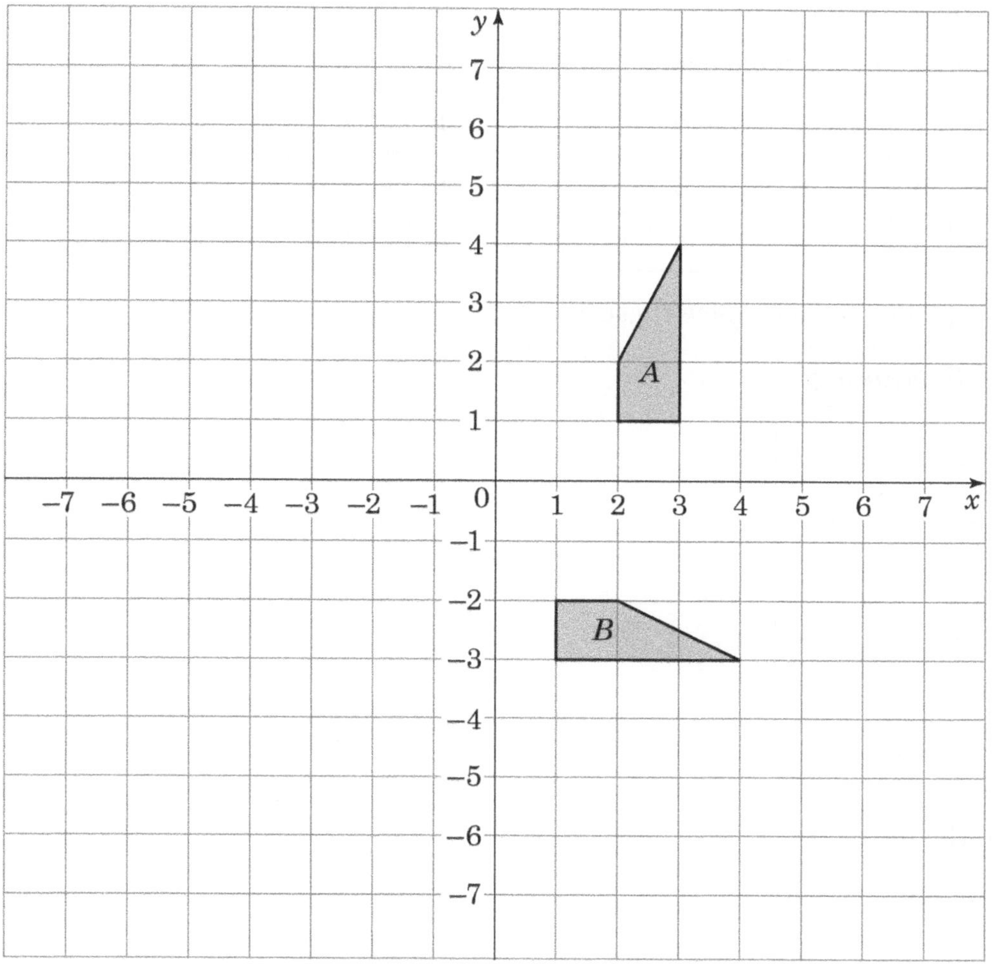

a) Describe fully the single transformation that maps shape A onto shape B.

____

____ **[3]**

b) On the grid, draw the image of:

i) shape A after a translation by vector $\begin{pmatrix} -4 \\ -3 \end{pmatrix}$; label this shape C **[2]**

ii) shape A after an enlargement by scale factor 2, centre (4, 1); label this shape D. **[2]**

25. a) Amit rolls a six-sided dice 60 times.

The results are shown in the table below.

4	4	3	1	5	3	2	5	3	1	5	1	3	4	1	5	2	5	5	4
1	6	2	6	5	2	6	4	5	3	4	5	6	4	6	3	5	1	1	2
5	5	4	2	5	6	5	1	5	6	3	6	3	2	1	2	4	6	4	5

Complete the frequency table to show these results.

Outcome	Frequency
1	
2	8
3	
4	10
5	16
6	9

[1]

b) i) Write down the mode. ____________ **[1]**

ii) Find the median.

____________ **[2]**

Amit rolls a different dice.

He draws a frequency chart to show the results.

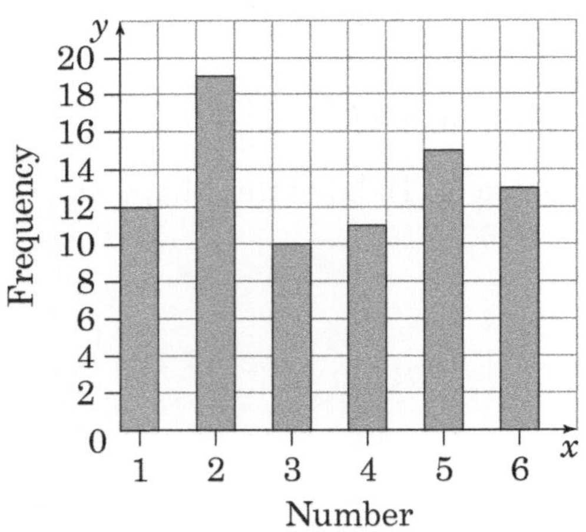

c) How many times did he roll the second dice? ____________ **[2]**

Examination-style Paper 3 (Calculator)

Paper 3 Calculator (Core)

2 hours

- You must answer **all** questions.
- You may use a calculator.
- You will need geometrical instruments.
- All necessary working should be shown clearly.
- Unless otherwise stated, give non-exact answers to 3 significant figures. For angles in degrees, give non-exact answers to 1 decimal place.
- The total mark for this paper is 80.

List of formulae

Area, A, of triangle, base b, height h	$A = \frac{1}{2}bh$
Area, A, of circle of radius r	$A = \pi r^2$
Circumference, C, of circle of radius r	$C = 2\pi r$
Curved surface area, A, of cylinder of radius r, height h	$A = 2\pi rh$
Volume, V, of cylinder of radius r, height h	$V = \pi r^2 h$
Curved surface area, A, of cone of radius r, sloping edge l	$A = \pi rl$
Volume, V, of cone of radius r, height h	$V = \frac{1}{3}\pi r^2 h$
Surface area, A, of sphere of radius r	$A = 4\pi r^2$
Volume, V, of sphere of radius r	$V = \frac{4}{3}\pi r^3$
Volume, V, of prism, cross-sectional area A, length l	$V = Al$
Volume, V, of pyramid, base area A, height h	$V = \frac{1}{3}Ah$

Calculators can be used on this paper.

1. $P = \frac{x}{2}(4x - 5)$

Find the value of P when $x = 16$.

$P =$ ____________ **[2]**

2. Calculate

$$\frac{16.4 + 21.07}{\sqrt{19.63} - 0.72}$$

Give your answer correct to 2 decimal places.

____________ **[2]**

3. Lexi changes $560 into euros.

The exchange rate is $1 = 0.864 euros.

Calculate the amount of euros that Lexi receives.

____________ **[1]**

4. The distance of the Earth to the Sun is approximately 149 600 000 kilometres.

a) Write this number in standard form.

____________ **[2]**

The average distance between the Earth and Mercury is 51.5% of the distance from the Earth to the Sun.

b) Calculate the average distance between the Earth and Mercury.

____________ **[2]**

5. Insert one pair of brackets to make this statement correct.

$$3 \times 9 - 4 + 7 = 22$$

[1]

6. An iron bar is 170 cm long.

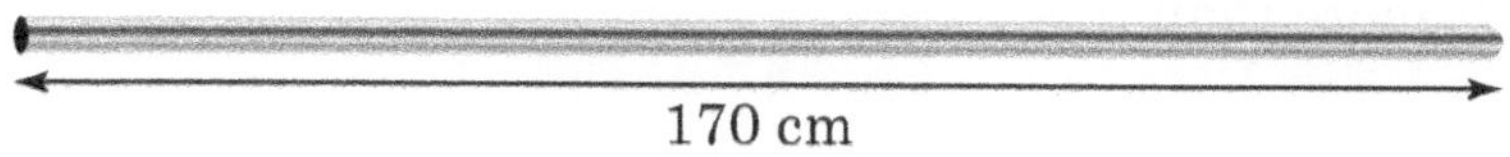

NOT TO SCALE

It has to be cut into equal lengths of $4\frac{1}{2}$ cm.

How many of these lengths can be cut from the bar?

______________ **[1]**

7.

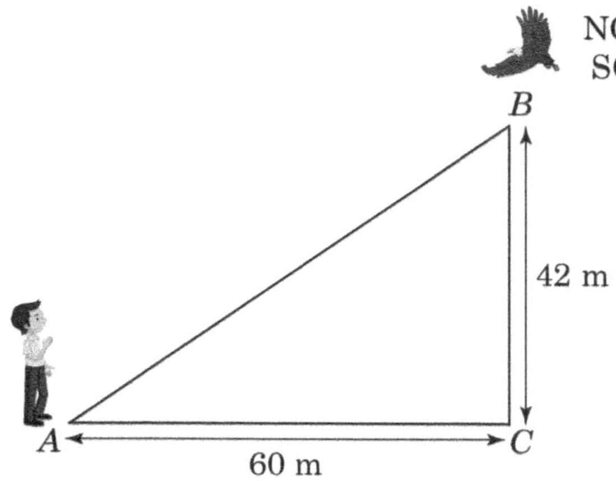

NOT TO SCALE

A bird, *B*, is hovering 42 m vertically above a point *C* on the ground.

A boy is standing at point *A*, which is 60 m horizontally from *C*.

a) Calculate the distance, *AB*, of the boy from the bird.

Give your answer correct to 1 decimal place.

AB = ______________ m **[2]**

b) Calculate the angle *BAC*. Give your answer to the nearest degree.

Angle *BAC* = ______________ **[2]**

8. Write down the reciprocal of 6.

______________ **[1]**

9. The temperature in Calgary, Canada on 13 December is −13.2 °C.

The temperature on 23 July is 34.7 °C more than this.

Calculate the temperature on 23 July.

_______________ °C **[1]**

10. A triangle has sides 5 cm, 6 cm and 8 cm.

Using a ruler and pair of compasses only, construct the triangle.

Show all of your construction lines.

[2]

11.

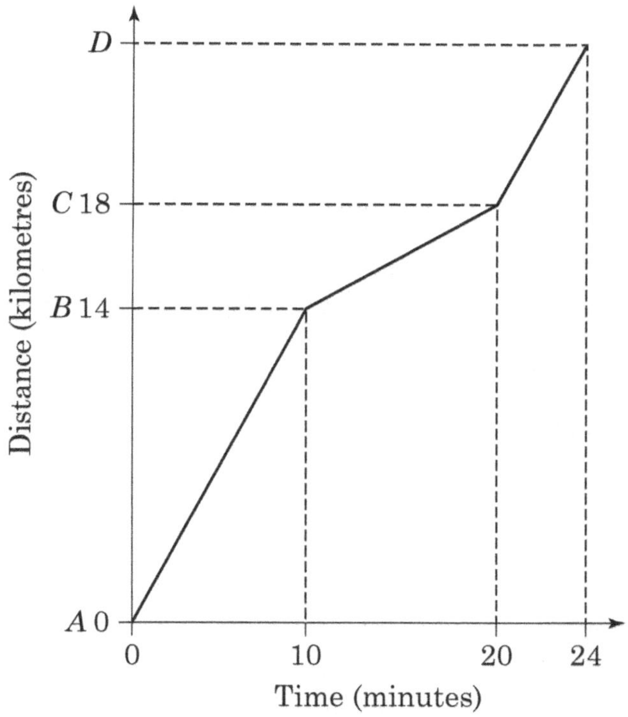

NOT TO SCALE

The graph shows Paula's motorcycle ride.

Starting at A, she rode 14 km to B at a constant speed.

Between B and C, she rode through a built-up area.

Between C and D, she rode at her original speed.

a) For how many minutes was Paula riding through the built-up area?

_______________ min **[1]**

b) At what speed did she ride through the built-up area?

Give your answer in:

i) kilometres per minute

_______________ km/min **[1]**

ii) kilometres per hour.

_______________ km/h **[1]**

c) What is the total distance from A to D?

_______________ km **[2]**

12. Write down all of the factors of 78.

____________________ **[2]**

13. The mean height of 8 male basketball players is 1.82 m.

The mean height of 6 female basketball players is 1.74 m.

Calculate the mean height of all 14 basketball players.

Give your answer to the nearest centimetre.

____________________ m **[3]**

14. Bryson walks at a speed of 1.4 metres per second.

Write this speed in kilometres per hour.

____________________ km/h **[2]**

15. Find the circumference of a circle of radius 4.2 cm.

Write down your answer:

a) exactly as it appears on your calculator

____________________ **[1]**

b) correct to the nearest centimetre.

____________________ **[1]**

16. Manuela earns \$450 per week.

She divides her earnings between rent, savings, entertainments and groceries.

a) She spends $\frac{2}{9}$ of her weekly earnings on rent.

Calculate her weekly rent.

________________ **[2]**

b) She divides the rest of her earnings between savings, entertainment and groceries in the ratio $10:11:14$.

Calculate the amount she spends on entertainment.

________________ **[2]**

c) She saves the same amount every week.

Calculate how much she saves in a year.

________________ **[2]**

d) At the end of the year, she invests her savings from the whole year in an account which pays 4% per year compound interest.

How much is the investment worth after two years?

________________ **[3]**

17. a) Calculate the area of a circle with a radius of 3.4 cm.

_______________ cm^2 **[2]**

b) Tennis balls are packed into a cylindrical can with radius 3.4 cm and height x cm.

The volume of a single can is $741\,cm^3$.

Calculate the value of x.

_______________ **[2]**

c)

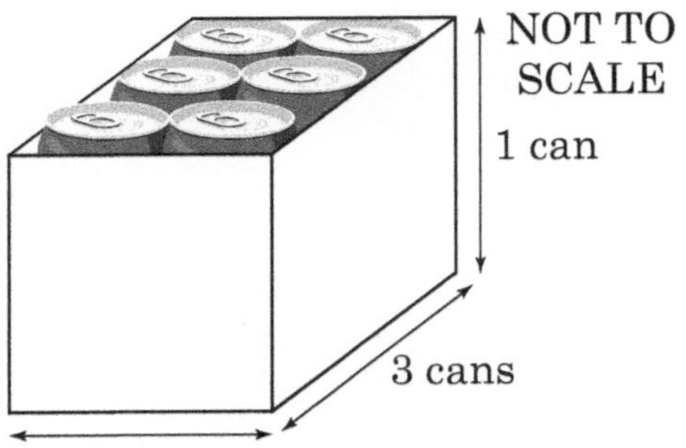

Six cans of tennis balls fit exactly into a box, as shown.

Write down the length, width and height of the box.

_______________ cm

_______________ cm

_______________ cm **[3]**

d) Calculate the percentage of the volume of the box occupied by the cans.

_______________ **[5]**

18. Wilma scores x marks in her driving theory test.

Manjit scores 3 marks more than Wilma.

Timothy scores twice as many marks as Wilma.

a) Write down a fully simplified expression for the total number of marks scored by Wilma, Manjit and Timothy. Write your expression in terms of x.

______________________ **[4]**

b) The mean mark scored by the three people was 33.

Find the value of x.

______________________ **[3]**

c) What mark did Timothy score?

______________________ **[1]**

d) A person took the test and scored 54 out of 70.

Work out this person's score as a percentage.

______________________ **[2]**

19. A solid metal pyramid has a square base with side length 25 cm and a perpendicular height of 12 cm.

a) Calculate the volume of metal in the pyramid.

_______________ cm^3 **[2]**

The pyramid is melted down and made into spheres, each of radius 2 cm.

b) Calculate the number of whole spheres that can be made.

_______________ **[3]**

The spheres are to be coated with a rust-resistant paint.

Each tin of paint can cover 500 cm^2.

c) Calculate the number of tins that are needed.

_______________ **[3]**

20. The stem-and-leaf diagram shows the scores of each of 21 golfers.

6	5	6	6	7			
7	1	1	1	4	8	9	9
8	0	2	4	5	6	9	
9	2	5	5	6			

Key: 6 | 5 represents a score of 65

a) Write down the range of scores.

_______________ **[2]**

b) Write down the modal score.

_______________ **[1]**

c) A 22nd golfer records a score of 79.

Explain how your answer to part **b)** will change.

____

____ **[1]**

21. The diagram shows the net of a cuboid.

NOT TO SCALE

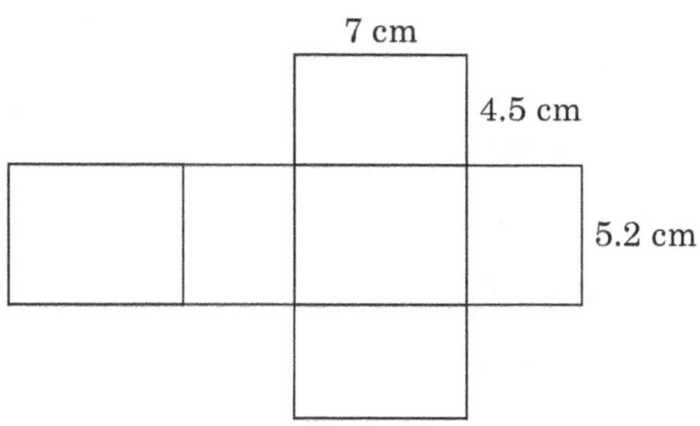

a) Calculate the volume of the cuboid.

_______________ cm^3 **[2]**

b) Calculate the total surface area of the cuboid.

_______________ cm^2 **[2]**

Answers and mark schemes

Chapter 1 Extension worksheet answers

1. 2, 3, 5, 7, 11, 13, 17, 19, 23, 29, 31, 37, 41, 43, 47

2. 73 only

3. **a)** **(i)** 20 **(ii)** 120 **b)** **(i)** 5 **(ii)** 550 **c)** **(i)** 30 **(ii)** 600

d) **(i)** 4 **(ii)** 1600 **e)** **(i)** 10 **(ii)** 8000 **f)** **(i)** 25 **(ii)** 1500

Chapter 2 Extension worksheet answers

1. 25, 36 (square numbers)

2. 26, 37 (add 9, add 11, etc.)

3. 15, 21 (triangular numbers)

4. 17, 23 (add 5, add 6, etc.)

5. 62, 80 (add 16, add 18, etc.)

6. 16, 32 (double each time)

7. 81, 243 (triple each time)

8. 48, 96 (double each time)

9. 0, 3, 8, 15, 24

10. 2, 4, 8, 16, 32

11. 4, 7, 12, 19, 28

12. 2, 6, 12, 20, 30

13. 1, 4, 16, 64, 256

14. 1, 8, 27, 64, 125

15. −1, 1, 5, 13, 29

16. 3, 9, 17, 27, 39

17. −1, 1, 7, 17, 31

18. 27, 81, 243, 729, 2187

19. **a)** 1, 1, 2, 3, 5, 8, etc.

b) 1, 4, 10, 20, 35, etc.

c) 1, 5, 12, 22, 35, etc.

Chapter 3 Extension worksheet answers

1. Three more squares are required to make the pattern have rotational symmetry of order 2.

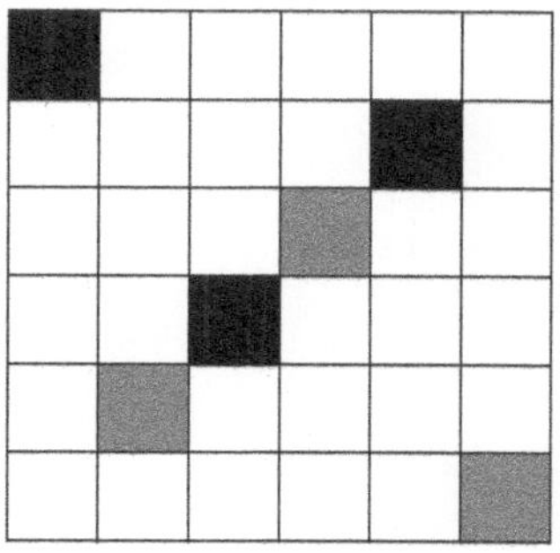

Nine more squares are required to make the pattern have rotational symmetry of order 4.

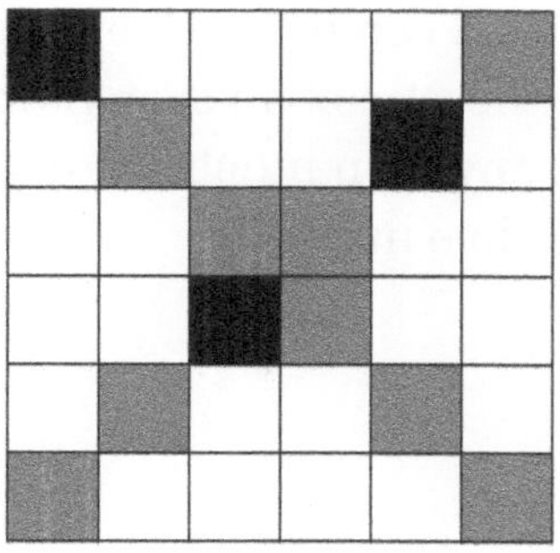

To shade the maximum number of squares, you could do this:

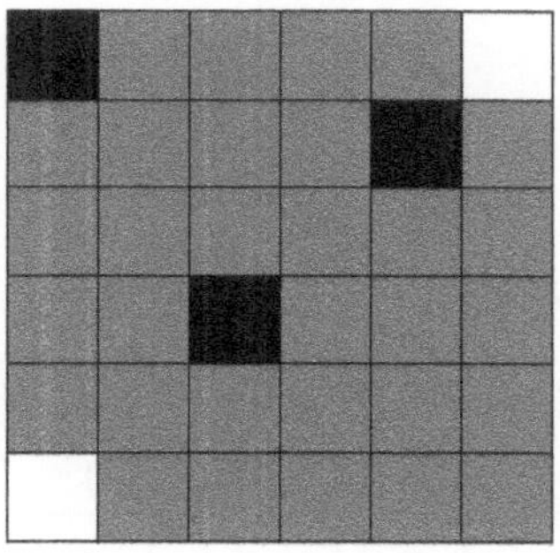

2. Students' own answers.

Chapter 4 Extension worksheet answers

General questions

1. The first number that can be made in three ways is 22.

$22 = 3 + 19$

$22 = 5 + 17$

$22 = 11 + 11$

The first number that can be made in four ways is 34.

$34 = 3 + 31$

$34 = 5 + 29$

$34 = 11 + 23$

$34 = 17 + 17$

The first number that can be made in five ways is 54.

$54 = 7 + 47$

$54 = 11 + 43$

$54 = 13 + 41$

$54 = 17 + 37$

$54 = 23 + 31$

2. The reason why any odd numbers greater than 5 can be made by adding three primes is because all odd numbers are the sum of an even number and 3. Then, if Goldbach's conjecture is true, any even number greater than 2 can be made by adding together two prime numbers.

This means that:

$7 = 4 + 3 = 2 + 2 + 3$

$9 = 6 + 3 = 3 + 3 + 3$

$11 = 8 + 3 = 3 + 5 + 3$

and so on.

Investigation questions

Students' own answers.

Chapter 5 Extension worksheet answers

1.

Weight (kg)	Frequency	Mid-intervalvalue(MIV)	MIV × frequency
$40 < w \leqslant 50$	2	45	90
$50 < w \leqslant 60$	8	55	440
$60 < w \leqslant 70$	7	65	455
$70 < w \leqslant 80$	3	75	225
	Total = 20		Total = 1210

estimate of the mean = total of MIV × frequency ÷ total frequency
$= 1210 \div 20 = 60.5$ kg

2.

Time (s)	Frequency	Mid-intervalvalue(MIV)	MIV × frequency
$0 < t \leqslant 20$	5	10	50
$20 < t \leqslant 30$	14	25	350
$30 < t \leqslant 40$	8	35	280
$40 < t \leqslant 60$	3	50	150
	Total = 30		Total = 830

estimate of the mean = total of MIV × frequency ÷ total frequency
$= 830 \div 30 = 27.7$ s (3 s.f.)

3.

Height (cm)	Frequency	Mid-intervalvalue(MIV)	MIV × frequency
$80 < h \leqslant 90$	6	85	510
$90 < h \leqslant 100$	10	95	950
$100 < h \leqslant 120$	18	110	1980
$120 < h \leqslant 150$	9	135	1215
$150 < h \leqslant 200$	7	175	1225
	Total = 50		Total = 5880

estimate of the mean = total of MIV × frequency ÷ total frequency
= 5880 ÷ 50 = 117.6 cm

Chapter 6 Extension worksheet answers

There is a limit to what you can earn if the interest is spread out through the year.

The following table illustrates this.

Unlike in the previous example, the totals have not been rounded to the nearest cent every time the interest has been calculated.

Times per year	Percentage each time	Unrounded Total	Rounded total
1	100.00	2	\$2.00
2	50.00	2.25	\$2.25
4	25.00	2.44140625	\$2.44
6	16.666...	2.521626372	\$2.52
8	12.50	2.565784514	\$2.57
10	10.00	2.59374246	\$2.59
12	8.333...	2.61303529	\$2.61
15	6.666...	2.632878718	\$2.63
20	5.00	2.653297705	\$2.65
24	4.1666...	2.663731258	\$2.66
48	2.08333...	2.690496599	\$2.69
72	1.3888...	2.699642059	\$2.70
100	1.00	2.704813829	\$2.70
150	0.666...	2.709275911	\$2.71
200	0.50	2.711517123	\$2.71
250	0.40	2.712865123	\$2.71
300	0.333...	2.713765158	\$2.71
365	0.2397...	2.714567482	\$2.71

As you can see, the maximum amount you can earn in one year this way from investing \$1 in an account like this is \$2.71.

This is long before you get to thinking about the difference between calculating the interest every day or every hour.

The maximum you can earn is the limit of the expression $1+\frac{1}{n}^{n}$ as n approaches infinity.

This is the definition of the mathematical constant e, which is approximately 2.71828.

This is why the maximum you can earn is \$2.71.

Chapter 7 Extension worksheet answers

If h_2 is defined to be the height of the cone that is removed from the top of the bigger cone, then

$$\frac{R_1}{R_2} = \frac{h + h_2}{h_2}$$

Cross multiplying gives

$$h_2R_1 = hR_2 + h_2R_2$$
$$h_2R_1 - h_2R_2 = hR_2$$
$$h_2 = \frac{hR_2}{R_1 - R_2}$$

Applying this gives the following solutions.

1. $R_1 = 4$ cm, $R_2 = 1$ cm, $h = 3$ cm

$$h_2 = \frac{hR_2}{R_1 - R_2} = \frac{3 \times 1}{4 - 1} = \frac{3}{3} = 1$$

volume of big cone – volume of small cone

$$= \frac{1}{3}\pi\left(R_1^2(h + h_2) - R_2^2h_2\right)$$
$$= \frac{1}{3}\pi\left(4^2(3 + 1) - 1^2 \times 1\right)$$
$$= \frac{1}{3}\pi(16 \times 4 - 1)$$
$$= 21\pi$$
$$= 66.0 \text{ cm}^3 \text{ (3 s.f.)}$$

2. $R_1 = 8$ cm, $R_2 = 4$ cm, $h = 6$ cm

$$h_2 = \frac{hR_2}{R_1 - R_2} = \frac{6 \times 4}{8 - 4} = \frac{24}{4} = 6$$

volume of big cone – volume of small cone

$$= \frac{1}{3}\pi\left(R_1^2(h + h_2) - R_2^2h_2\right)$$
$$= \frac{1}{3}\pi\left(8^2(6 + 6) - 4^2 \times 6\right)$$
$$= \frac{1}{3}\pi(64 \times 12 - 96)$$
$$= 224\pi$$
$$= 704 \text{ cm}^3 \text{ (3 s.f.)}$$

3. $R_1 = 10$ cm, $R_2 = 2.5$ cm, $h = 2$ cm

$$h_2 = \frac{hR_2}{R_1 - R_2} = \frac{2 \times 2.5}{10 - 2.5} = \frac{5}{7.5} = \frac{2}{3}$$

volume of big cone − volume of small cone

$$= \frac{1}{3}\pi\left(R_1^{\ 2}(h + h_2) - R_2^{\ 2}h_2\right)$$

$$= \frac{1}{3}\pi \ \ 10^2 \ \ 2 + \frac{2}{3} \ \ - 2.5^2 \times \frac{2}{3}$$

$$= \frac{1}{3}\pi \ \ 100 \times 2\frac{2}{3} - \frac{25}{6}$$

$$= \frac{262.5}{3}\pi$$

$= 275$ cm^3 (3 s.f.)

4. $R_1 = 20$ cm, $R_2 = 4$ cm, $h = 7$ cm

$$h_2 = \frac{hR_2}{R_1 - R_2} = \frac{7 \times 4}{20 - 4} = \frac{28}{16} = \frac{7}{4}$$

volume of big cone − volume of small cone

$$= \frac{1}{3}\pi\left(R_1^{\ 2}(h + h_2) - R_2^{\ 2}h_2\right)$$

$$= \frac{1}{3}\pi \ \ 20^2 \ \ 7 + \frac{7}{4} \ \ - 4^2 \times \frac{7}{4}$$

$$= \frac{1}{3}\pi \ \ 400 \times 8\frac{3}{4} - 28$$

$$= \frac{3472}{3}\pi$$

$= 3640$ cm^3 (3 s.f.)

5. $R_1 = 11.5$ cm, $R_2 = 5.75$ cm, $h = 4.5$ cm

$$h_2 = \frac{hR_2}{R_1 - R_2} = \frac{4.5 \times 5.75}{11.5 - 5.75} = \frac{25.875}{5.75} = 4.5$$

volume of big cone − volume of small cone

$$= \frac{1}{3}\pi\left(R_1^{\ 2}(h + h_2) - R_2^{\ 2}h_2\right)$$

$$= \frac{1}{3}\pi\left(11.5^2(4.5 + 4.5) - 5.75^2 \times 4.5\right)$$

$$= \frac{1}{3}\pi\left(132.25 \times 9 - 148.78125\right)$$

$$= \frac{11109}{32}\pi$$

$= 1090$ cm^3 (3 s.f.)

Chapter 8 Extension worksheet answers

1. −29, −10, −3, −2, −1, 6, 25

2. 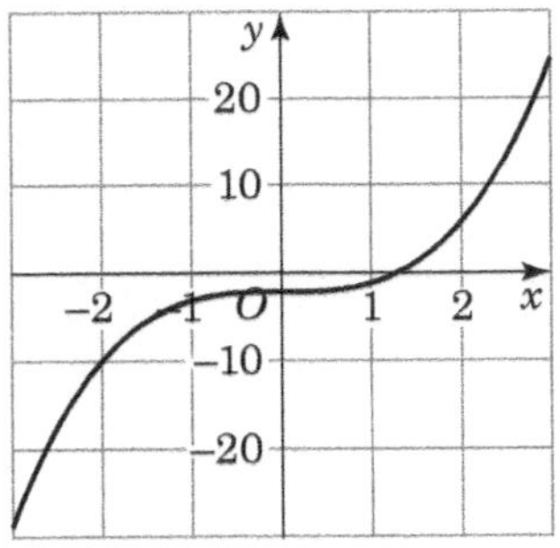

3. $x = 1$ to 1.5

4. $x = 1.75$ to 2.25

5. $y = x$, $x = 1.25$ to 1.75

6. **a)** $x = -1.17$ **b)** $x = 1.70$ **c)** $x = -1.36$

7. It is undefined; 10, 2.5, 2, 1, 0.5, 0.33

8. 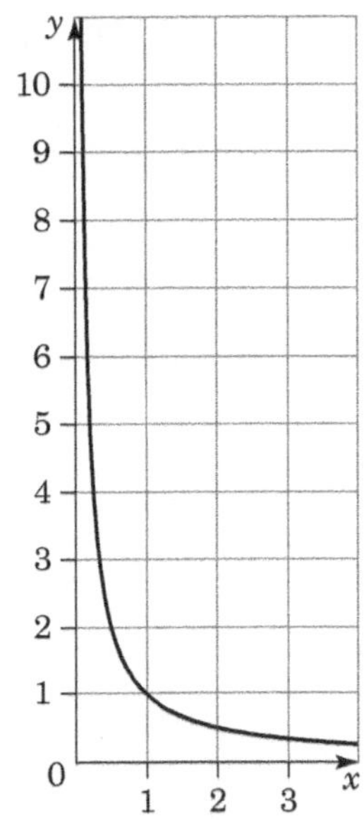

9. $x = 0.33$

10. $x = 0.67$

11. $x = 0.62$

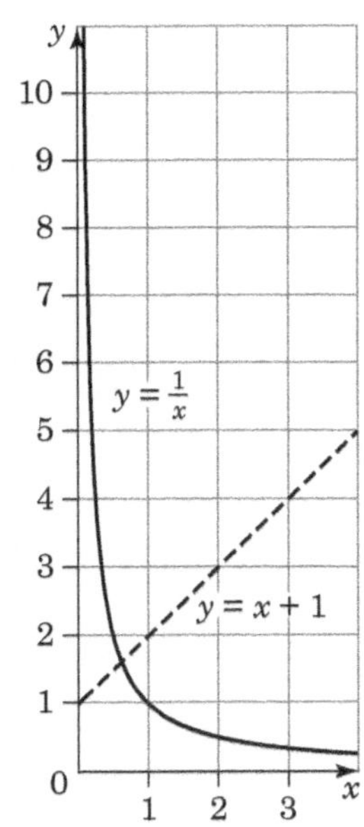

Chapter 9 Extension worksheet answers

1.

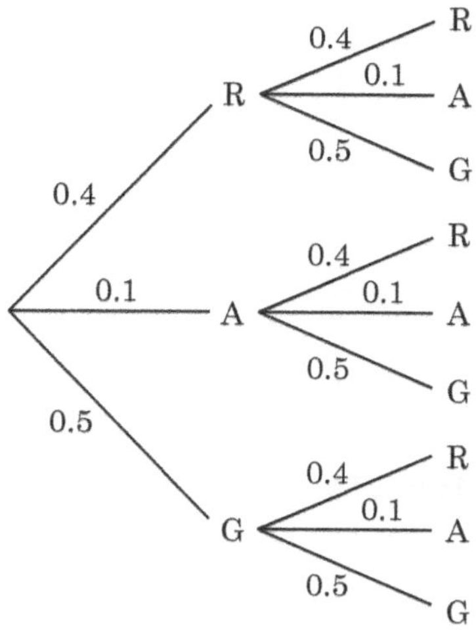

a) P(RR) = $0.4 \times 0.4 = 0.16$

b) P(AA) = $0.1 \times 0.1 = 0.01$

P(GG) = $0.5 \times 0.5 = 0.25$

P(same colour)
= P(RR) + P(AA) + P(GG)
= $0.16 + 0.01 + 0.25 = 0.42$

2.

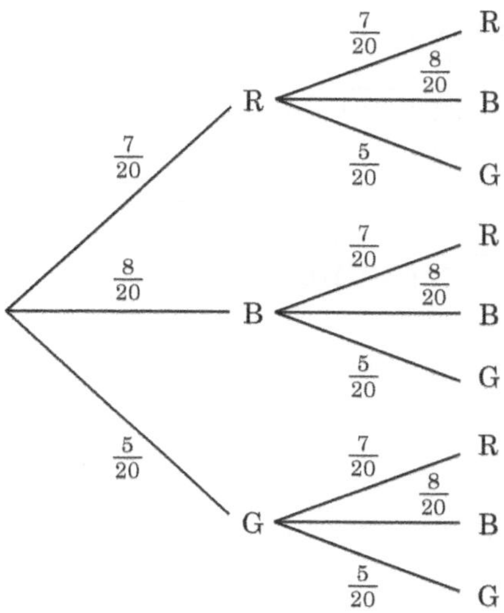

P(different colours)
= 1 – P (two the same colour)

$$P(RR) = \frac{7}{20} \times \frac{7}{20} = \frac{49}{400}$$

$$P(BB) = \frac{8}{20} \times \frac{8}{20} = \frac{64}{400}$$

$$P(GG) = \frac{5}{20} \times \frac{5}{20} = \frac{25}{400}$$

P(two the same colour)

$$= \frac{49}{400} + \frac{64}{400} + \frac{25}{400} = \frac{138}{400} = 0.345$$

P(different colours) = $1 - 0.345$
= 0.655

3.

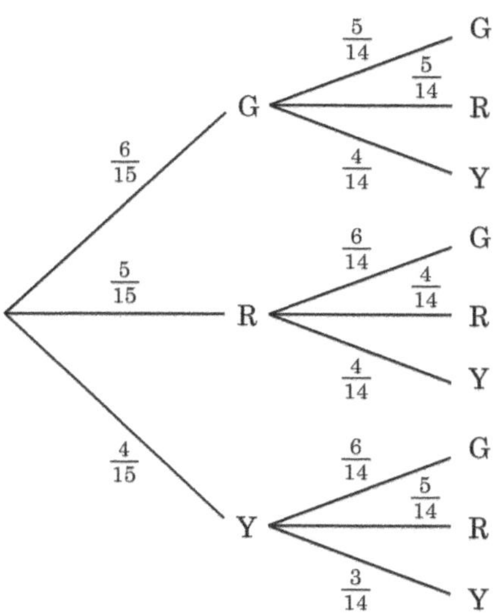

$$P(GG) = \frac{6}{15} \times \frac{5}{14} = \frac{30}{210}$$

$$P(RR) = \frac{5}{15} \times \frac{4}{14} = \frac{20}{210}$$

$$P(YY) = \frac{4}{15} \times \frac{3}{14} = \frac{12}{210}$$

P(two the same colour)

$$= \frac{30}{210} + \frac{20}{210} + \frac{12}{210} = \frac{62}{210}$$

$$= \frac{31}{105} \approx 0.295$$

Chapter 10 Extension worksheet answers

1. $x = 3, y = 0$ or $x = -3, y = 3$
2. $x = 3, y = 3$ or $x = -3, y = -7$
3. $x = 1, y = 2$ or $x = -1, y = 6$
4. $x = 3, y = 4$ or $x = -3, y = -8$
5. $x = 5, y = 0$ or $x = -5, y = 15$
6. $x = 2, y = 0$ or $x = -2, y = 2$

Chapter 11 Extension worksheet answers

All answers are given to 3 s.f.

1. $\sqrt{6^2 + 4^2 + 3^2} = \sqrt{36 + 16 + 9} = \sqrt{61} = 7.81\,\text{cm}$
2. $\sqrt{7^2 + 3^2 + 2^2} = \sqrt{49 + 9 + 4} = \sqrt{62} = 7.87\,\text{cm}$
3. $\sqrt{10^2 + 5^2 + 6.5^2} = \sqrt{100 + 25 + 42.25} = \sqrt{167.25} = 12.9\,\text{cm}$
4. $\sqrt{25^2 + 12.2^2 + 11^2} = \sqrt{625 + 148.84 + 121} = \sqrt{894.84} = 29.9\,\text{cm}$
5. $\sqrt{8.6^2 + 4.8^2 + 2.3^2} = \sqrt{73.96 + 23.04 + 5.29} = \sqrt{102.29} = 10.1\,\text{cm}$
6. $a = \sqrt{d^2 - b^2 - c^2}$
7. $a = \sqrt{18^2 - 7^2 - 6^2} = \sqrt{239} = 15.5\,\text{cm}$

Multiple choice revision test 1 answers

1.	D	2.	B	3.	A	4.	A
5.	C	6.	D	7.	B	8.	D
9.	C	10.	B	11.	D	12.	C

Multiple choice revision test 2 answers

1.	C	2.	C	3.	B	4.	B
5.	D	6.	C	7.	B	8.	C
9.	D	10.	A	11.	A	12.	C
13.	A	14.	C	15.	B		

Multiple choice revision test 3 answers

1.	C	2.	D	3.	B	4.	C
5.	D	6.	B	7.	D	8.	C
9.	B	10.	A	11.	A	12.	C

Multiple choice revision test 4 answers

1.	A	2.	D	3.	C	4.	B
5.	A	6.	D	7.	B	8.	D
9.	B	10.	D	11.	C	12.	B
13.	B	14.	C				

Examination-style Paper 1 (Non-calculator) mark scheme

Question	Answer	Marks	Partial marks
1. a)	655	1	
1. b)	Eight hundred and forty five	1	
2.	Lower bound: 235 Upper bound: 245	1 1	
3.	$k = \frac{a - c}{b}$	2	M1 for subtracting c
4. a)	Triangular prism	1	
4. b)	9	1	
5. a)	68	2	M1 for $16 \div 4 \times 17$
5. b)	17	1	
6.	100	1	
7.	$3a(a - 2b)$	2	B1 for either a or 3 taken out
8. a)	41	1	
8. b)	4	1	
8. c)	$\frac{1}{25}$ (accept 0.04)	2	B1 for answer $\frac{1}{n}$ or 25
9. a)	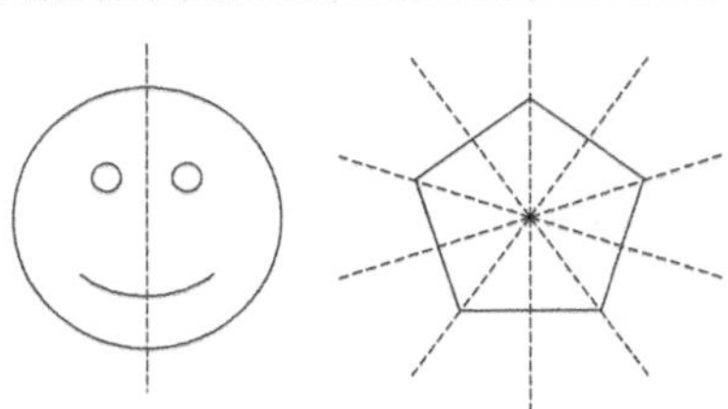	2	1 mark for each
9. b)	5	1	
10.	4	4	M1 for $\frac{1}{2} \times 12 \times 5$ M1 for $54 -$ *their* 30 M1 for *their* $24 \div 12 \times 2$
11.	Positive correlation	1	
12.	70°	2	M1 for $ADC = 30°$ or $CDF = 40°$
13.	-1	2	M1 for $5x = k$ oe
14. a)	-4	1	
14. b)	$y = -4x + 8$	1	
15. a)	13	1	
15. b)	31	1	
15. c)	$3n + 1$	2	B1 for $3n$
16. a)	5	1	
16. b)	$-7 \leqslant x < 4$	2	B1 for $x \geqslant -7$ or $x < 4$
17.	36	2	M1 for $30 \div 5$
18.	$15 - 6x$	2	M1 for $4x - 10x \pm 15$

<table>
<tr><td>19.</td><td>24</td><td>1</td><td></td></tr>
<tr><td>20.</td><td>$x = 3$
$y = -2$</td><td>3</td><td>M1 for correct method to eliminate one variable
A1 for one correct answer</td></tr>
<tr><td>21. a)</td><td><table><tr><td>x</td><td>−3</td><td>−2</td><td>−1</td><td>0</td><td>1</td><td>2</td><td>3</td><td>4</td><td>5</td></tr><tr><td>y</td><td>−16</td><td>−8</td><td>−2</td><td>2</td><td>4</td><td>4</td><td>2</td><td>−2</td><td>−8</td></tr></table></td><td>3</td><td>B2 for 3 correct
B1 for 1 or 2 correct</td></tr>
<tr><td>21. b)</td><td>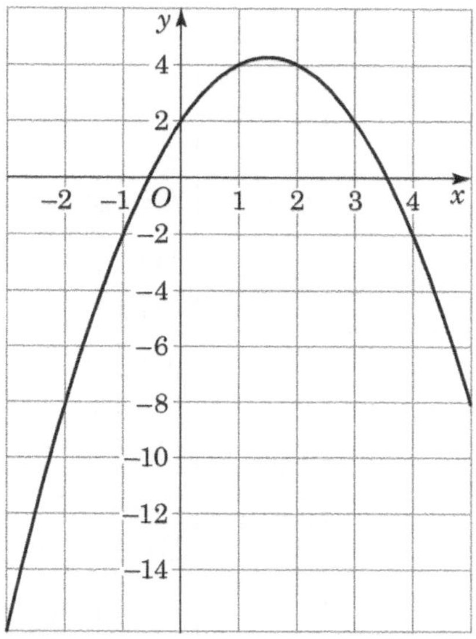
</td><td>4</td><td>B2 for 7 or 8 points correct (ft their table)
B1 for 5 or 6 points correct (ft their table)
B1 (independent) for smooth curve through their points</td></tr>
<tr><td>21. c)</td><td>−0.6
3.6</td><td>2</td><td>1 mark for each</td></tr>
<tr><td>21. d) i)</td><td>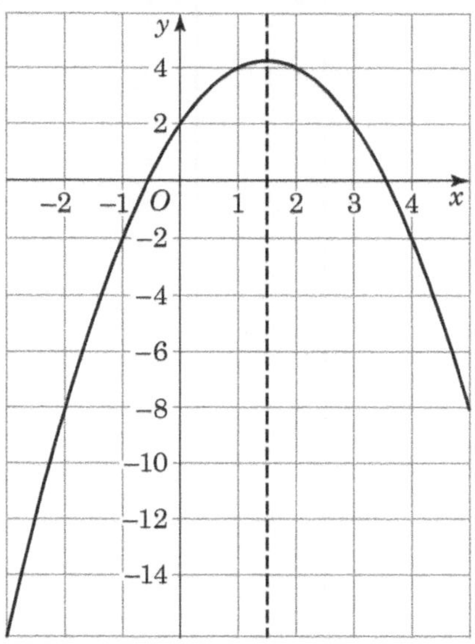
</td><td>1</td><td></td></tr>
<tr><td>21. d) ii)</td><td>$x = 1.5$</td><td>1</td><td></td></tr>
</table>

22. a)	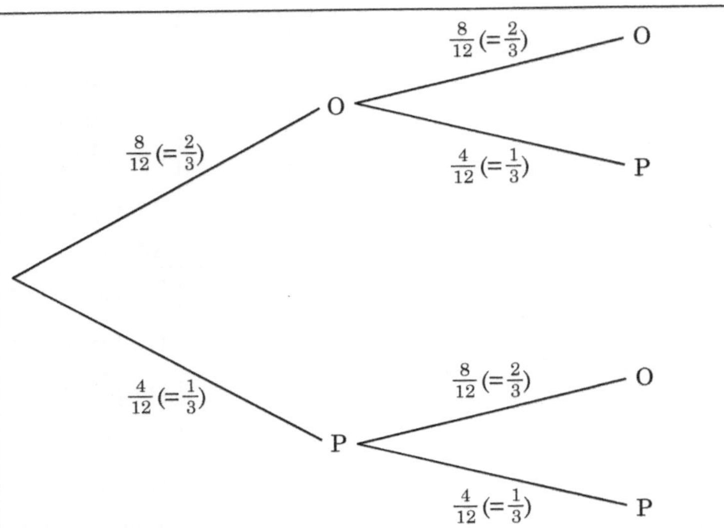	2	B1 for first level branches correct
22. b)	$\frac{5}{9}$	2	Accept unsimplified M1 for either P(OO) or P(PP)
23. a)	108°	2	M1 for (5 – 2) × 180 or finding exterior angle
23. b)	72°	2	M1 for 360 ÷ 5
23. c)	54°	1	ft *their* (a) and (b)
24. a)	Rotation, 90° clockwise about (0, 0)	3	B1 for rotation B1 for 90° clockwise B1 for (0, 0) (accept 'origin')
24. b) i)	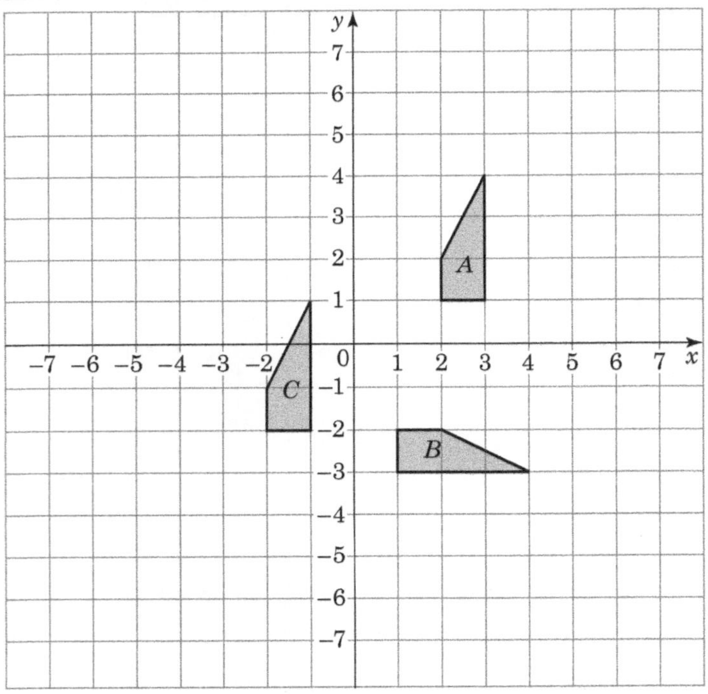	2	B1 for translation by $\begin{pmatrix}k\\-3\end{pmatrix}$ or $\begin{pmatrix}-4\\k\end{pmatrix}$

24. b) ii)	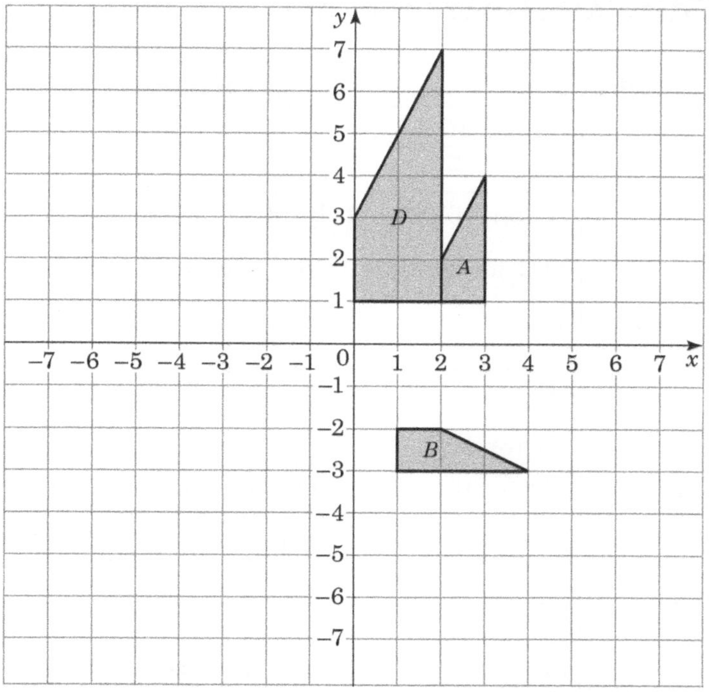	2	B1 for enlargement sf 2 in the wrong position
25. a)	9; 8	1	
25. b) i)	5	1	
25. b) ii)	4	2	M1 for identifying 30.5th data value
25. c)	80	2	M1 for evidence of reading at least 4 frequencies correctly

Examination-style Paper 3 (Calculator) mark scheme

Question	Answer	Marks	Partial marks
1.	472	2	M1 for $\frac{16}{2}(4 \times 16 - 5)$ or better
2.	10.10	2	B1 for 10.0981 to 10.0982 or 10.1 SC1 for *their* 10.098… seen and rounded correctly to 2 dp
3.	483.84	1	
4. a)	1.496×10^8	2	B1 for 1.496 B1 for 10^8
4. b)	77 044 000	2	M1 for use of 0.515 as multiplier or finding 1% Accept answer in standard form (7.7044×10^7)
5.	$3 \times (9 - 4) + 7 = 22$	1	
6.	37	1	
7. a)	73.2	2	M1 for use of Pythagoras' theorem with given lengths
7. b)	35°	2	M1 for use of tangent with given lengths [alt: use of sin or cos with *their* (a)]
8.	$\frac{1}{6}$	1	
9.	21.5	1	
10.	Correct ruled triangle with construction arcs	2	B1 for correct triangle with incorrect/no arcs or for two correct arcs
11. a)	10	1	
11. b) i)	0.4	1	
11. b) ii)	24	1	
11. c)	23.6	2	M1 for finding the speed from A to B and multiplying by 4 to find the distance from C to D
12.	1, 2, 3, 6, 13, 26, 39, 78	2	B1 for six or seven correct and no extras or all correct and one extra
13.	1.79	3	M2 for $\frac{1.82 \times 8 + 1.74 \times 6}{8 + 6}$ M1 for 1.82×8 or 1.74×6
14.	5.04	2	M1 for $\frac{1.4 \times 60 \times 60}{1000}$
15. a)	26.38937829 or $\frac{42}{5} \neq$	1	
15. b)	26	1	
16. a)	\$100	2	M1 for dividing by 9
16. b)	\$110	2	M1 for dividing *their* \$350 by 35
16. c)	\$5200	2	M1 for *their* savings $\times$ 12

16. d)	\$5624.32	3	M1 for use of 1.04 as multiplier A1 for amount after 1 year (\$5408) or M2 for use of 5200×1.04^2
17. a)	36.3	2	M1 for use of πr^2
17. b)	20.4	2	M1 for use of formula for volume of a cylinder and *attempt* to rearrange (before or after substitution)
17. c)	20.4; 13.6; 20.4	3	1 mark for each correct ft *their* (b)
17. d)	78.6	5	M1 for attempt to find volume of box A1 for volume of box ft *their* (c) M1 for 6 times volume of cylinder M1 for dividing A1 for percentage ft *their* volume of box if consistent with (c)
18. a)	$4x + 3$	4	B1 for correct expression for Manjit B1 for correct expression for Timothy M1 for adding three expressions
18. b)	24	3	M1 for dividing *their* (a) by 3 or multiplying 33 by 3 M1 for attempt to solve resulting equation
18. c)	48	1	ft *their* x
18. d)	77.1	2	M1 for $54 \div 70$ oe
19. a)	2500	2	M1 for use of correct formula $\frac{1}{3} \times 5 \times 5 \times 12$
19. b)	74	3	M1 for volume of sphere $\frac{4}{3} \times \pi \times 2^3$ M1 for dividing *their* (a) by volume of sphere
19. c)	8	3	M1 for surface area of sphere $4 \times \pi \times 2^2$ M1 for *their* (b) multiplied by *their* surface area and dividing by 500
20. a)	31	2	M1 for $96 - 65$
20. b)	71	1	
20. c)	There will be two modes (71 and 79)	1	
21. a)	163.8	2	M1 for $7 \times 4.5 \times 5.2$
21. b)	182.6	2	M1 for 7×4.5 or 4.5×5.2 or 7×5.2